The Growing Gap Between Emerging Technologies and Legal-Ethical Oversight

The International Library of Ethics, Law and Technology

VOLUME 7

For further volumes:
http://www.springer.com/series/7761

Gary E. Marchant · Braden R. Allenby ·
Joseph R. Herkert
Editors

The Growing Gap Between Emerging Technologies and Legal-Ethical Oversight

The Pacing Problem

Foreword by Peter A. French

Springer

Editors
Gary E. Marchant
Arizona State University
College of Law
PO Box 877906
85287-7906 Tempe Arizona
USA
gary.marchant@asu.edu

Braden R. Allenby
Arizona State University
Ira A. Fulton School of Engineering
Department of Civil and Environmental Engineering
PO Box 875306
85287 Tempe Arizona
USA
braden.allenby@asu.edu

Joseph R. Herkert
School of Letters and Sciences
Arizona State University
Santa Catalina Hall 250 D
85212 Mesa Arizona
USA
joseph.herkert@asu.edu

ISSN 1875-0044 e-ISSN 1875-0036
ISBN 978-94-007-1355-0 e-ISBN 978-94-007-1356-7
DOI 10.1007/978-94-007-1356-7
Springer Dordrecht Heidelberg London New York

Library of Congress Control Number: 2011928677

Printed on acid-free paper

Springer is part of Springer Science+Business Media (www.springer.com)

Foreword

On November 4, 1811 an armed gang attacked the home of a master weaver in the village of Bulwell in the Midlands of England. The gang's mission was to destroy six weaving machines. They succeeded in smashing them all. They called themselves the followers of General Ludd (or King Ludd or Captain Ludd) and for a little over a year they terrorized the Midland counties of England, busting textile machinery. The Luddites were skilled weavers, artisans of clothes-making, many stockingers, being driven out of business by the use of automated looms operated by unskilled labor that produced goods, primarily mittens and stockings, at lower prices, of inferior quality, and at six times the quantity that the fruits of their skilled labor could fashion.

In about three months the Luddites destroyed 1,100 textile machines. The British government took action on the side of the new industrialists, passed a law making the penalty for destroying a machine death by hanging, and sent in the Army to enforce it. The law was widely interpreted as proclaiming that machines have a greater value than humans. Death sentences and other penalties were carried out when machine smashing Luddites were captured. At least 24 Luddites were hanged, 44 were transported to Australia, and 18 imprisoned. The damage to machinery in the Luddite forays was estimated at £100,000.

The Luddites claimed in their pamphlets that they were not against machinery or technology per se. Their opposition was focused on "machinery hurtful to commonality." By that, they seemed to mean that they opposed the use of technologies over which the general population, presumably through their local governments, had no control and that were in their view inimical to the collective well-being and interests of that population. For the skilled weavers in the Luddite bands the machinery that they attacked fit that description not only because it produced clothes inferior to the products of their handiwork, but because its use reduced their standards of living substantially while economically benefiting only well-heeled industrialists. Their rage against the machine was triggered by the threat of job losses, penury, and to the very existence of their trades.

The Luddites, of course, were utter failures. Technology moved on and at a pace that those institutions of society that should be concerned about positive and negative effects on "the commonality" have been unable or unwilling to match ever since. The Luddites disappeared as an organized body, but their legacy, their name, survives as a pejorative description of technophobes, and that is unfair to those skilled weavers that smashed mechanical looms in the early nineteenth century. There is something of value to be learned from the Luddites and it is not that technological progress must be stopped whatever the cost. It is that there may be a need to regularly evaluate our conception of moral permissibility and the legal apparatuses we employ to protect and defend the general interests of the community in order to insure that they are not stalled at the starting blocks when technological innovations capable of producing radical changes in communities and lifestyles are way ahead of them on the track, modifying virtually every aspect of communal and personal life. That is not to say that modifications are not welcome or that many are not significant improvements. Many are beneficial, but at the same time they create considerable shifts in the way the world must be perceived and in the conception of our individual and collective places in it. Think of the social scene before and after the automobile or the train. The wide-spread presence of the television sets in the homes of Americans radically altered neighborhood life in the 1950s, while in the 1960s the Vietnam War was not just somewhere in Southeast Asia, but, thanks to television, it was in most American living room every day.

The traditional moral theories with roots in the pre-industrial Enlightenment are typically trotted out by philosophers to confront what they perceive – generally with minimal understanding of the capacities of the technology, the science, and the engineering – to be imminent or potential assaults on the common good by the development, merchandising, or use of the products and processes of new technologies. For those theories to function usefully in real time in supporting regulatory decisions the number of unknown or unknowable variables relevant to their calculi must be minuscule. How, for example, can one decide what is the greatest good for the greatest number of people if the size of the impacted population is indeterminable, and it is not clear what the effects are until they actually happen? If we have learned anything in this area in the last half-century, it should be that predicting significant outcomes of the emerging technologies across populations is virtually impossible. Who could have predicted in the 1960s when the United States funded military research projects on distributed computer networks that a quarter of the earth's population now would be using the Internet in ways that have a profound impact on us, reshaping our lives, radically changing how we understand and interact with... everything?

We are living in a world that is profoundly different from the one in which the ethical theories with which we are most familiar, Enlightenment ethics, were invented, and in which many of our legal processes and policies were crafted. Our world is, in all of its aspects, including what it means to be human, contingent upon the outcomes of engineering/design projects. Of course, there has always been an element of the unknown, and the uncertain, involved in being human; in this sense the world in which we are living is not greatly different from the one depicted in the extant fragments of the ancient Greek philosopher Heraclitus. Two fragments give something of the flavor. Heraclitus writes[1]: "Time is a child moving counters in a game; the royal power is a child's." He seems to mean that the randomizing activity of a very young child moving the pieces of a game like chess without any attention paid to the constitutive and regulative rules of that game is an apt metaphor for the world in which we have to function. To extend the metaphor, it is virtually impossible to predict where the pieces will next land on the board, or whether they will be scattered off the board, or if the board itself will be in a jumble on the floor, if there is a floor. There are a number of famous Heraclitian fragments that deal with the impossibility of stepping twice in the same river. However, the last fragment in that vein goes further than maintaining that the river is flowing, so that the water into which one steps a second time cannot be the water into which one first stepped. "Into the same rivers we step and do not step. We exist and we do not exist," he wrote. It is not just that we cannot step twice in the same river; we are not the same persons from moment to moment. Both the river and we are constantly in a state of flux.

Were Heraclitus writing today, he might comment that what it is to be human, and a particular human, is contingent upon unanticipated impacts that interactions with other fluctuating elements of a mostly engineered environment may randomly produce. We live in the brave new world of the GRINN technologies – genetic engineering, robotics, information technology, neuroscience, and nanotechnology, and all manner of other technologies that transform the material and the mental spaces in which we function. Persons (or their minds and identities) now can be distributed over information systems and design platforms of various kinds, for example as avatars in *Second Life*, and, in many cases, actively engage in multiple realities. The Enlightenment moralists and those who designed our legal systems would, by and large, be confused, if not utterly lost, in our world. But then, so, in large measure, are most of us.

Classical utilitarianism, for example, may be formulated to say that an act is right, the thing we ought to do, if and only if, in the circumstances its consequences have a higher utility than any other thing we could have performed. Utility was defined in terms of happiness or pleasure. Hence, the utilitarian principle typically is stated as "Choose the action in the circumstances that will produce the greatest happiness for the greatest number of people." The maximization of two independent variables (pleasure and population) is involved in determining what ethically we as a society ought to do. And to a large extent much of our regulatory tools are based on some sort of utilitarian basis: do what is best for the greatest number of people. But in many cases involving the emerging technologies of our era, those are unknown or unknowable variables. How then can utiliatarianism serve to guide us in

[1]The quotations attributed to Heraclitus are from *Heraclitus, The Complete Philosophical Fragments*, translated with commentary by William Harris at http://community.middlebury.edu/~harris/Philosophy/Heraclitus.html.

policymaking regarding those technologies? How can it help us decide what will be good for and what will be detrimental to society? And what if we also worry about sustainability and impacts on future generations? An ethical system for decision-making regarding public policies regulating the emerging technologies will not be much good if, at best, it were only retrospective, if it could only tell us what we should have done after an outcome of our actions has already been irretrievably set in motion or occurred. If we have virtually no way of accurately predicting what populations will be adversely or advantageously affected by the introduction of a new technological innovation, how can we use the utility calculus to tell us what we ought to do? Utilitarianism, in all of its forms, assumes a large number of constants in order to arrive at recommendations it can endorse. But none of the things it holds constant, including the humans whose happiness (or good or preferences) is to be maximized, are excludable from technological design. Humans themselves are now design spaces. Everything is in a Heraclitean flux.

That other great moral theory from the Enlightenment, Kantian deontology, fares no better. Kant's categorical imperative requires that we not decide on a course of action that we cannot universalize across the population of the moral community and that we treat all members of that community with dignity and respect, as ends and never as means only and that we not privilege ourselves over the interests of others, that we always act as both subject and sovereign in a kingdom of ends. Kant is primarily concerned not with outcomes, but with processes, not with how things turn out, but with the motives people have when they choose courses of action. An ethical person is to act always so that he or she could rationally will that the principle on which he or she is acting could become a universal law for all people. The difficulty is that in cases involving the new technologies we usually cannot say what effects they will have on populations, so we cannot be at all certain that we would not be using large numbers of people as means and not as ends were we to approve or disapprove of the introduction of this or that technological innovation or process, even with the best of intentions.

What we need, it would seem, is an ethics and a legal structure designed to respond to the contingencies and floating factors and variables that are commonplace in, indeed characteristic of, the technological potential of the Twenty-first Century. Ethicists need to be asking how much in human experience must be constant for ethical principles to be efficacious. If even the subjects of ethical principles and rules are alterable to suit conditions and whims, how do we decide what should and should not be done? Legal structures and procedures should be designed not to stymie technological innovation, but to insure that the Luddite principle regarding controlling what is "hurtful to commonality" is given sufficient attention as cases are adjudicated and policy is formulated, nationally and globally.

I do not know what form the requisite ethical systems and legal apparatuses to handle the problems and issues that are arising and that will continue to arise with respect to the emerging technologies should take. The task of designing the ethics and the legal response mechanisms that can keep pace with technology and science is a paramount challenge facing us.

Ethics and law have typically lagged far behind technological change. If we allow that lag time to increase, it will grow exponentially until both ethics and law will be realistically viewed as an irrelevant antique of a time long past and not fondly remembered. Cognizant of that likelihood, the Lincoln Center for Applied Ethics at Arizona State University, of which I am the Director, funded the first few years of what we called "The Pacing Project." Lincoln Professors Braden Allenby, Joseph Herkert, and Gary Marchant conceived of the project and have been directing it. This volume is a result of their opening the dialogue among ethicists, legal scholars, engineers, and technologists facing the challenge of trying to close the gap between the sprinting technologies and the plodding ethical and legal systems that are supposed to be concerned with regulating and recommending policy and procedures to minimize the effects of "machinery hurtful to commonality." The Lincoln Center for Applied Ethics is proud to be associated with this project.

Peter A. French

Acknowledgements

This research, writing and editing of this book was supported in part by a seed grant from the Lincoln Center of Applied Ethics at Arizona State University and a grant from the National Science Foundation (Award SES-0921806). The views expressed in this manuscript are those of the authors and do not necessarily represent the views of the Lincoln Center, the National Science Foundation, or U.S. Government.

Introduction: Why Law and Ethics Need to Keep Pace with Emerging Technologies

Andrew Askland

The evidence is unmistakable that knowledge in the sciences and various familiar and new technologies is accumulating in prodigious amounts. The accumulation is notable both for its quantity and for its variety. More noteworthy, the accelerating rate of this accumulation is staggering. Whereas the Enlightenment model had suggested a neat ordering of knowledge that might permit a good encyclopedia to circumscribe what constituted knowledge, now a more appropriate model is a swiftly moving target that continually accretes data and links that data to other data, a monumental Wikipedia continually updated with expert input. The linkages elude exact specification because the linkages connect with knowledge that is also continuously expanding. The expansions are not linear nor are they easily predicted and the potential synergies are confounding. It is like an expanding universe with more than the four familiar dimensions, a string theory account with exponential growth along each string. The technologies of recent origin, i.e., the "five horsemen" identified by Brad Allenby, viz., nanotechnology, biotechnology, robotics, information / communication technologies, and applied cognitive science, are making it increasingly obvious that, as Gary Marchant phrases it, "concurrent technological revolutions are rapidly transforming economic, social and personal domains, now and even more so in the imminent future."

The charge is that oversight and governance is not keeping pace with technological and scientific change. For example, efforts to regulate the impact of information technologies upon privacy often prove ineffectual because they are drafted to constrain technologies and practices that are continually changing and those changes escape the language of the rapidly outdated regulations. Information technologies evolve quickly and regulations often address a snapshot of a technology that is tracking quickly away from the conceptual scheme of the regulations. When technologies change so fast that consumers must accept that anything that they buy is already obsolete, how does the law stay apace? How do we evaluate the significance of the charge, i.e., what are the consequences if the legal system and ethics lag behind

A. Askland (✉)
Arizona State University, Tempe, AZ, USA
e-mail: sandy.askland@asu.edu

rapid changes in scientific and technological knowledge? Why does it matter if a gap opens between rapidly expanding knowledge and the law's grasp of that knowledge? These questions are important because the law serves a crucial social function that depends upon its broad applicability and its predictability and ethics can degenerate into empty platitudes when it is ignorant of the relevant facts in moral problems for which it seeks to provide guidance. Perhaps we should concede that the gap is inevitable and learn to accommodate it. Perhaps, as Scott McNeely advised *vis a vis* the loss of privacy, we need to "get over"[2] the puncturing of a popular bromide, in this case, a presumption that law should track developments in science and technology. On the other hand, it may be that the gap is a cause for alarm because it undermines the institutions and the practices of both law and science. The argument here is that the law does need to keep pace with rapid changes in knowledge and that the costs of failure are substantial. That is hardly a startling response to the gap, but it is worth drawing out, at least briefly, the impact of a substantial divide between technological/scientific knowledge and the ability of legal and ethical systems to competently grasp that knowledge.

Most democratic forms of government are divided into three familiar branches: judicial, legislative, and executive, and it is useful to describe how each performs its tasks in an ideal situation. Actual conditions fall considerably short of ideal, but the shortcomings of practice are nonetheless measured against the standard of an unrealizable model. Judges are in fact subject to biases that reflect their social and economic backgrounds. Legislators are often primarily concerned with their re-election and placating popular discontents rather than marshalling the best facts and strongest arguments to devise bills that address matters of paramount public importance. Regulators are often captured by "public choice" strategies (Buchanan and Tullock 1962) to advance their own interests rather than the purposes identified for the regulations that they promulgate and enforce.

These practical shortcomings undermine the credibility and effectiveness of governance institutions and programs, but it is a mistake to focus exclusively upon the disappointments of applied politics. There is much to value and build upon in our systems of government. Judges apply established rules to resolve the particular problems argued before them. They draw out the rationales of rules (and precedent) to identify the principles that explain the rules (and older holdings) in order to apply those principles to new variations of fact. Legislators represent the interests of the electorate and endeavor to express those interests and the principles that undergird those interests by enacting laws that connect those interests / principles to pressing social and economic challenges. Regulators fashion rules that elaborate upon principles articulated in governing legislation to efficiently guide the conduct of parties subject to the regulatory regime devised by the legislature.

These functions can be described more fully, but a cursory summary is sufficient to make the point that each branch is involved with facts and principles and,

[2]Scott McNeely is the co-founder of Sun Microsystems; he said "You have zero privacy; get over it." This famous quote was attributed to him by Stephen Manes in "Private lives? Not ours" in PC World 18 (6):312, 18 April 2000.

moreover, that facts inform the identification of suitable principles and principles cannot be well applied where the facts to which they are applied are misunderstood. The law cannot articulate suitable principles without a grasp of the facts that are relevant to those principles. Similarly, the law cannot apply principles without an understanding of the facts that constitute the problem. Thus, if an important purpose of law is to promote fair and beneficial outcomes for problems that are posed in specific factual contexts (applying known and endorsed legal norms), then law must maintain a facility with the facts of the problems that are or will be posed.

If legislation and regulation seek to provide legal guidance before the fact and adjudication seeks to sort out difficulties after the fact, then legal actions will certainly disappoint unless they are appropriately informed about the relevant facts. If we expect law to competently address the safety of new technologies, then the new technologies must be sufficiently well understood that their implications for safety can be evaluated and managed. If we expect law to sort out the value of new technologies in order to determine the cost of its accessibility, or the appropriateness of public subsidies, or the availability of immunities/safe havens, then the new technologies must be sufficiently well understood to enable a probative evaluation of those costs and benefits. There are many reasons to avoid resort to legal process, e.g., the costs of generating appropriate legislation or of taking cases to trial, the delays occasioned by these efforts, the uncertain outcomes, etc., but when disputes arise, as they inevitably do, the legal system is the ultimate means for resolving those disputes. Disputes are, in fact, often resolved without resort to legal intervention, but those resolutions are devised in the shadow of the legal interventions that will occur unless the parties can amicably resolve their differences. An informed grasp of the relevant facts and principles promotes efficient decisions about how law will likely be applied and what terms of negotiated compromise are appropriate.

A legal structure combines explicit law, implicit practice, and a broader cultural framework within which the law is situated, which in turn implicates ethical, social and economic considerations. Law is the means by which we attempt to resolve conflicts and law operates more predictably when it commands the relevant facts about the underlying conflicts. Law must be ready to competently address these conflicts and the greater the potential impact of the conflict, the greater the need to anticipate resort to legal intervention. Where there is a social impact, there is cause for competence to assess the costs/benefits of that impact. These impacts are easiest to identify when they resemble familiar legal categories, e.g., torts and contracts, but they also encompass legal issues that are less easily categorized, e.g., the definition of personhood or duties owed to non-human subjects or the international externalities of technological development.

One might argue that law should accept an intermediate resolution of the challenge of new science and emerging technologies and wait for those developments to cross an impact threshold before focusing attention upon them. It may be that law should stay abreast of scientific and technological advances to the extent that contractual disputes occur among parties, or compensation is sought for tortious conduct, or boundaries need to be identified for intellectual property claims, or other familiar legal issues arise in the context of scientific and technological research.

One might argue that, aside from these frictions, law need not worry about a divide between scientific and legal knowledge. As research nears application, the law should take note, but only as it anticipates the extension of familiar jurisdictional definitions to new subject matter. Law need not worry about its ignorance of the subject matter or the methodologies of the research in its formative stages save for focused concerns, e.g., about informed consent for human subjects or intrinsically dangerous subject matter (and these concerns do require considerable attention).

As the European Science Foundation recently reported regards stem cell research, "Most scientists are overwhelmed when they review all relevant regulations and guidance covering the storage, use and disposal of secretions, organs and tissue for regenerative research." (ESF 2010) This approach carves out a substantial area of concern for legal inquiry, but preserves a default for scientific research that lies beyond the current competence of law. Is it a suitable compromise to allow law to assure that certain important conventions are satisfied, e.g., institutional review boards to approve human and animal studies and funding agencies to control research that involves dangerous chemicals, etc., during the trail and error of research, and wait until the late stages of scientific and technological advancements before worrying about their contents, e.g., let the appropriate regulatory body address the approval of a product or service when it is offered for public consumption? Does this approach provide sufficient protection for the public interest and is it a preferred, and more easily obtainable, objective for law's ambition to keep pace with science?

An initial response to this proposed compromise is based upon prudence. It will be easier for law to accommodate applied science if it closely tracks theoretical science. The learning curve to master the intricacies of an applied technology will be better handled if law follows the early movements along that curve. Waiting until shortly before a technology is introduced to grapple with its ambitions and potential impacts may unnecessarily delay the introduction and cause uncertainties that can be considerably reduced with an earlier familiarity. Indeed, this first response blends into a second. An earlier familiarity with the blossoming technology might affect the course of its development so that it is more readily accepted and approved at its introduction. That point inspires this rebuttal: should emerging technologies be affected in the course of their development by currently governing legal conventions? As new technologies, they likely press hard on those conventions. These concerns are relevant because technologies often become "locked in" – that is, connected to other economic and technological systems in such a way that significant change is either difficult or impossible after their adoption. As Carl Mitcham phrases it, technology can "transform social structures in ways that tend factually to predetermine their uses." (Mitcham 1994, p. 273) Once "lock in" occurs, it may be difficult for the law to prescribe alternative paths even if they were obviously preferable when the technology was forming. As Andrew Feenberg points out, "Design is only controversial when it is in flux." Once the conflict is resolved, technological and legal standards become embedded in stable code that shapes our daily practices and frames our perception of the value of those practices. (Feenberg 1999, p. 96)

Applied cognitive science suggests a model of human behavior which differs significantly from the model that governs most theories of criminal and tort responsibility. Biotechnologies suggest radically different conceptions of health care and medical malpractice and mortality. Other emerging technologies entail similar reorientations of familiar practices. These emerging technologies portend dramatic changes to the legal conventions that currently prevail and it might seem that subjecting those technologies to review by soon obsolescent conventions will be counterproductive, i.e., the threatened conventions might thwart the normal course of development for new technologies. The obverse phrasing is that to the extent that emerging technologies portend such dramatic changes, they ought to be tracked with considerable care and attention lest the law be unprepared for those changes. If the changes are impending, then law will benefit from ample notice to accommodate them. If the changes are unavoidable, but their form is unspecified, then law ought to be able to negotiate about that form.

This negotiation role surely should not be limited to law, but as law reflects important ethical, social and economic values, it ought to be included among the negotiators. It is because emerging technologies will radically transform our culture and our polity that the parameters of those potential changes ought to be examined by many diverse parties. At a minimum we expect the institutions that currently resolve most social and economic disputes to actively engage with the pending changes.

Our institutions of public learning, our ethical leaders, and our journalistic enterprises, among others, should also be engaged, but it is imperative that legislatures track the new knowledge to assure that the public interest is appropriately served (and create a means for regulatory oversight where that is required) and that the judiciary similarly track this knowledge so that is prepared to adjudicate disputes. (The judicial role is emphasized in the United States by the adoption of the Daubert standard for the admissibility of expert scientific evidence at trail. This standard imposes a gatekeeper responsibility upon the trial judge to assure that proffered expert scientific evidence relies upon a valid scientific methodology. This responsibility has driven many judges to educate themselves about emerging technologies and the foundations of science.)

This broader engagement also benefits science and technology. It opens their work to perspectives free from the tunnel vision of a specific discipline or research agenda. The intense concentration of sustained and particularized research can generate a myopic focus that blinkers the researcher and dulls an appreciation for the broader implications of the work. It benefits the research and the researcher to periodically explain the work and consider its effects in larger contexts than an isolated study. In short, research can be improved by regular engagement with those whose lives will be affected by its outcomes. The results of research are better appreciated and its implications better acknowledged when the work is periodically vetted by parties and interests outside the scientific and technological communities. A backlash against the research is less likely when the work is perceived to be embedded in the polity that supports the work and whose interests and values are being advanced by the work. Science and technology are, as Daniel Sarewitz points out, social enterprises and it is entirely reasonable to expect that their focus and progress should be justified routinely to the polity in which the research occurs.

A simple statement of the function of law is that it aims to influence behavior, either by discouraging particular behaviors with penalties, or incentivizing other behaviors with subsidies or non-interference. It is easier for law to influence behavior when it is reinforcing behavior rather than contravening it. Most criminal law is an elaboration upon basic civil norms, e.g., proscribing assaults, murder, thefts, fraudulent acts, etc. Much that has been written about the relationship between law and morality focuses upon the role of broadly shared norms of behavior in grounding the legitimacy of particular laws and legal systems. (Fuller 1960; Raz 1978; Greeenawalt 1989)

Of course, many laws are stipulative or conventional, e.g., which side of the road we drive on. Many regulations are justified by the public good promoted by consistency where many individualized behaviors would be onerously chaotic. In either case, law can more productively influence behavior when it is informed about the purposes and consequences of that behavior. Law can better decide when it should not attempt to influence behavior, e.g., where such efforts are inappropriate or inefficient, when it has a confident grasp of what is at stake. If the norms for science are opaque to law, if the benefits and costs of particular technological endeavors are secreted among a parochial corps of investigators, or are otherwise indecipherable or badly misapprehended, then it is likely that law will not phrase its legislation, its regulations or its judgments to suitably coordinate the priorities of science and technology with generally prevailing social norms. There will be persisting mismatch between, on the one hand, basic civil norms and the public good served by standardization and, on the other, the purposes and practices of science and the emerging technologies. Instead of a healthy engagement between science/technology and law which would inform each about the other's priorities and practices, there will be structural ignorance that is disruptive and costly. Cooperation is difficult, especially between contrasting cultures, (Snow 1959; Goldberg 1994), especially when those cultures are tracking rapidly away from one another, but the costs of a failure to cooperate can be disastrous.

The concern that science and technology are outpacing law recalls past concerns that a steadily increasing number of statutes was inexorably encumbering legal systems. Many statutes address conditions that change and the statutes are not timely revisited to gauge their appropriateness for the changed conditions. Guido Calabresi addressed this mismatch of statutes with changed circumstances and argued against reliance upon administrative agencies to update statutes to comport their original intentions with changed circumstances. (Calabresi 1982) He thought that administrative agencies would lack a sufficiently broad perspective on the changed circumstances and would not renovate statutes frequently enough. He was also critical of renovation by administrative agencies because they were not sufficiently majoritarian and would lack the requisite adherence to principles and consistency. They were also subject to capture by vested interests, notably the corporate parties affected by the regulations, and also were conflicted by their own preferences and a related bias against modifications to the status quo. He argued instead for judicial authority to rephrase anachronistic statutes so that the statutes would harmonize with the prevailing legal fabric. He thought that judges possessed the skills and perspective to identify the "underlying values of a people." (p. 98) Judges, informed by the trajectory of statutory enactments, by scholarly criticisms, and by "the gravitational pull of deep constitutional principles," (p. 99) can evaluate new situations and revise old statutes to promote "conformity with a complex legal landscape." (p. 100) Calabresi was not advocating rash judicial activism, but instead sought a means to revitalize statutes that had been rendered ineffectual by changed circumstances. Whatever principles they were written to promote had been overtaken by changes that rendered them, without modification, poor fits and incompetent standards. However, a judicial authority to renovate anachronistic statutes is inconsistent with the role of judges in most civil law jurisdictions and likely strains what many people, skeptical about judicial overreaching, would accept in common law jurisdictions.

Accelerating advances in science and technology exacerbate the problem of unrevised statutes, but the crux of the problem is the same: a dissymmetry between law and newly arising facts because the law has not anticipated these new facts and attempts to govern them with an antiquated grasp of their meaning. Reprising the description of law as the matching of facts and principles, the challenge for law is to be keenly attentive to new facts in order to fashion principles that will sort out the challenges that they present. Law requires a competent grasp of those new facts to better recognize where its statutes (and precedents) require revision to best preserve what the society that it serves seeks to preserve in this hyper-technological world. In the best circumstances this will be difficult because technological evolution is unpredictable and it renders contingent many assumptions that are unreflectively regarded as fixed, and thus raises difficult challenges to the mission of pacing law with science and technology.

Given this view that law should strive to keep pace with science and technology, it is worthwhile to examine existing legal structures to identify how they might better accommodate new scientific and technological knowledge. It is also worthwhile to consider alternative legal structures and forms that might better accommodate that new knowledge than current legal structures. Finally, while law is the immediate regulator of much science and technology, ethical values often provide the foundation for both legal and broader societal responses to emerging technologies. Thus, both legal and ethical frameworks and systems must keep pace with rapidly evolving science and technologies.

There is a lengthy history of scientists serving as consultants in the legal process, whether in an official or unofficial capacity. Sheila Jasanoff has identified two common paradigms for the use of scientific input, a 'technocratic' approach, which makes "scientists the primary validators of policies with high technological content," and a 'democratic' approach, which relies upon "broad public participation as an antidote to abuses of expert authority." She instead recommends a 'negotiated' model for scientific inputs in order to "harness the collective expertise of the scientific community so as to advance the public interest." (Jasanoff, p. vii, 250) The extremes to be avoided are naïve deference to science and technology as complete solutions to social and political problems and blind reliance upon the popular will as the final authority in matters of fact and scientific theory. The challenge is to bring accurate accounts of emerging technologies to the public and their representatives to facilitate competent debate and appropriately informed laws.

This volume is an effort to promote the availability of scientific knowledge to legal actors who serve the pubic interest by considering the challenges presented by new scientific and technological knowledge and proposing various strategies that might better enable the legal and ethical frameworks to cope with and manage that knowledge. The first part of this book describes the "pacing problem." The second part explores some of the dynamics of the oversight challenge posed by emerging technologies. The third part provides a "toolbox" of possible solutions to help address the pacing problem.

Brad Allenby opens the "pacing problem" part with a chapter that focuses on the "systems" impacts of emerging technologies. He uses the example of the railroad to demonstrate the power of a new technological system (as a Kondratiev wave of innovation) to transform a society, provoking "profound and unpredictable institutional, organization, economic, cultural and political change." "[T]echnology of any significance tends to be profoundly destabilizing ... and thus, of course, to the degree it is so, it will generate substantial and potentially powerful opposition." He then describes the NBRIC (nano, bio, robotic, info and communication) technologies and their potential to similarly disrupt currently prevailing norms and practices. However, "it is not just that each NBRIC system is powerful; it is that they are combining in unexpected ways that are both beyond any single technological domain, and very potent." Allenby elaborates upon these synergistic effects with specific examples and argues that the emergence of these technologies "dramatically affects our usual assumptions about stability and cultural frameworks along three dimensions: complexity, contingency and accelerating change." He concludes that an ability to perceive and understand these dimensions of emerging technologies can permit us to "interact with them to achieve more desirable outcomes and trajectories" and "to develop institutional mechanisms, including legal and regulatory tools that engage with the systems in ways that are both productive and predictable."

Gary Marchant provides detailed evidence of the accelerated pace of developments in science and technology over the past several decades. Moore's Law and Monsanto's Law are examples of consistent exponential growth in computing power and biological knowledge, respectively. Similar productivity impacts have been identified for DNA synthesis and sequencing and for the resolution of neuronal features of computed tomography (CT) brain scanning. Internet connectivity has grown rapidly for many years, as have nanotechnology related patents. Marchant then provides examples of the failure of the existing legal framework "based on a static rather than a dynamic view of society and technology" to cope with these rapid advances. The U.S. Congress, for example, linked attainment requirements and ozone standards without allowing for changes to the ozone standards which would make unaltered enforcement of the attainment standards absurd. This mismatch is particularly striking because the Environmental Protection Agency (EPA) is required by the same statute to periodically upgrade the ozone standard and thus the absurdity would have been anticipated by anyone paying serious attention to the prospect of advances in scientific knowledge. Marchant also points out that the reaction times of legal institutions have slowed down in recent decades exacerbating the pacing problem, and he provides evidence for why that slowing down has occurred in legislatures, regulatory agencies and the courts.

There is a widespread awareness of the "pacing problem" and Marchant cites observations by various experts from multiple fields expressing their judgment that law is struggling to keep pace with new scientific developments. He concludes with a list of suggestions for devising more flexible and adaptive regulatory approaches to avoid or minimize the pacing problem: expedited rule-making; self-regulation (or "cooperative" regulation); issue specific statutes; courts with specialized subject matter jurisdiction; sunset clauses; periodic mandatory program reviews; free-standing independent institutions with specific issue foci; adaptive management strategies that use feedback to reorient their policies and practices; and principles-based regulation (rather than detailed prescriptive rules).

Joseph Herkert explicitly poses ethical questions that are latent in various other chapters. He explores whether emerging technologies require new ethical concepts or merely expand the scope of existing ethical concepts. He uses humanoid robotics and pervasive computing to frame his response. Human robotics are insinuating themselves into modern, technological societies in various commercial and military capacities. The increased use and reliance upon these increasingly sophisticated robots presses ethical worries beyond consumer safety and products liability to more fundamental ethical categories such as moral agency, free will and human identity. Pervasive computing is the convergence of advances in fields such as microelectronics, materials science, solid state physics, nanotechnology, radio frequency identification (RFID), wireless communications, and global positioning systems (GPS) and can affect "every facet of our lives, including our homes, workplaces, schools, businesses, and entertainment venues." These two technologies share with other emerging technologies a few characteristics which make them especial challenges to ethics: embeddedness; unlimited reach; an engineering of the mind and the body; and specificity.

Herkert points to engineering ethics as a useful source of perspective for these challenges and describes in particular a grid scheme that cross references various domains for engineering ethics as either macro-ethics or micro-ethics and either engineering practice or scientific research. He is attracted to such new formulations because there is a growing sentiment, at the very least within engineering ethics, that traditional ethical concepts have proven inadequate in general and in particular when confronted with emerging technologies. Herkert notes that promoters of the emerging technologies seem willing to abandon traditional ethical concepts, such as the primary role of human agency, which might impede the progress of technology, while critics of the emerging technologies tend to focus on the process of ethical deliberation, both in terms of timeliness and participation. He identifies a middle ground where moral imagination and preventive ethics are important themes that can be developed to devise new ethical concepts and frameworks that can meet the task of pending and future ethical challenges.

The second part of the book, focusing on oversight dynamics, opens with David Rejeski's chapter addressing public policy on the technological frontier. He notes that there are a number of reasons why public policy makers encounter difficulties on this frontier: its novelty; cognition biases of the policy makers; framing that distorts issues and debates; intractable problems with inadequate resources assigned to them; and many known unknowns. To confront these issues, Rejeski suggests that we rephrase the governing metaphors to deemphasize an "assessment and regulation" paradigm, with its "interminably long process of issue identification, analysis, recommendations, and implementation" to an emphasis on co-evolution. He also recommends an embedded early warning system approach to promote reflexive and anticipatory governance. Another useful (open-source) tool would be an evolving list of known unknowns to help reduce the likelihood of surprises where possible scenarios can be considered before they occur. Rather than keying on best practices, Rejeski instead focuses on bad practices in order to collect and manage information about them and channel that information into solutions. He also recommends that research scientists and engineers be appropriately trained so that they provide oversight for the research enterprises in which they are engaged rather then relying exclusively upon social scientists and ethicists from outside that enterprise. Finally, Rejeski recommends that we develop and implement learning strategies that focus upon learning from mistakes instead of obsessing about the impossible task of avoiding all mistakes. Advances in computation and rapid prototyping systems permit learning and innovation through experimentation that promotes better solutions informed by many tested hypotheses. Rejeski sums up these recommendations as efforts to rephrase learning about emerging technologies as co-evolution rather than an effort to run faster on the technology treadmill.

Deborah Johnson uses software agent technology to explore an anticipatory ethics approach to rapidly developing technologies. This approach engages with the ethical implications of a technology in its early stages of development in order to influence that development. The early engagement applies the insight of various Science and Technology Studies (STS) scholars that technological development is fluid and contingent because it is social negotiated and constituted. The software agent technology example, which involves complex systems designed to operate independently of their human designers, is instructive because these agents have been described by some commentators as autonomous and this characterization collides with traditional moral notions and practices of accountability. Johnson examines the strengths and weaknesses of an argument for a moral ontology for software agents that keeps them tethered to those who design and deploy them. Her analysis of software agents supports the conclusion that moral notions and practices do influence technology and that they can more intentionally and effectively address the pacing problem by affecting the future development of emerging technologies.

Lyria Bennett Moses considers the extent to which *sui generis* rules are an effective approach for dealing with the "pacing problem." The *sui generis* approach is often recommended where there is pressure to enact new laws for a new technology, or where uncertainty arises about the application of law in new contexts, or where the law may not apply as intended in the new context. She provides several examples where *sui generis* rules were adopted in response to technological change and describes the rationales for the *sui generis* approach in each case. She identifies and evaluates several potential disadvantages of the *sui generis* approach: the failure to cover sufficient ground (the completeness problem); the administrative costs of maintaining multiple legal regimes (the administrative cost problem); the tendency of *sui generis* rules to assume a temporary technological framework (the problem of technological change); and the potential for narrowly defined rules to unfairly benefit narrowly defined groups (the political problem). Moses then describes the pros and cons of *sui generis* rules, concluding that the decision on whether or not to adopt such an approach may be technology and context-specific and needs to be carefully and openly considered on a case-by-case basis.

Daniel Sarewitz identifies the context for governing technological change as the dilemma created by a commitment to pluralism, participation and openness in our governance structures, on the one hand, and the enormous transformational power of technology and technological systems, "a power that often seems at once inscrutable, unconscious, overwhelming, and autonomous," on the other. He describes how the challenge of this dilemma has been largely deflected: "the pursuit of technological transformation is largely exempted from formal democratic processes of eliciting value preferences and adjudicating value disputes about desired future states." He offers an alternative resolution for the dilemma, namely a technological assessment (TA) where the pace and direction of knowledge and its applications are directed by human choice, where those choosers encompass a range of socio-economic, cultural and political components, and where social settings interact with techno-scientific advances to create evolving outcomes that reflect the values of the widely encompassed decision-makers. Sarewitz offers the example of the real time technology assessment (RTTA) project pursued by the Center for Nanotechnology and Society at Arizona State University that attempts to exemplify the alternative approach by building reflexivity into the research process.

The third part of this volume describes an assortment of possible solutions to the "pacing problem." Brian Rappert describes his experience with codes of conduct as a possible means of addressing the challenge of keeping pace with science and technology. Codes of conduct encompass a variety of aims, drafters and target audiences, but are generally an attempted form of self-regulation. Rappert adjudges that, if our standard for assessing whether codes are working is their effects on guiding the behavior of practitioners, then this aim has largely been unrealized. Codes have tended to codify existing practices rather than establishing new standards that require changed behavior. Rappert counters that the process of deliberating about codes has helped track developments in science and technology by helping to build shared agendas and enabling coordinated initiatives and in that sense codes are working if we recalibrate our standard of review. Raising awareness about important topics, fostering ethical reflection about emerging issues, clarifying responsibilities and increasing public confidence are also valuable effects of the code drafting process. Rappert rehearses the history of efforts in the United Kingdom to draft codes for biological weapons to prevent the destructive use of life science research to support his evaluation.

Kenneth Abbott focuses on international action to coordinate national law and policy in response to scientific and technological innovation. International action can steer national law and policy toward greater uniformity, and also towards greater efficiency, legitimacy and a public interest orientation in form and content, and, moreover, facilitate speedier responses, especially in states with limited regulatory capacity. He provides a general institutional framework for international coordination, steering, and facilitation, with two key elements: a "framework convention" and a set of international institutions and procedures that would operate under the authority of that convention. He proposes arrangements that would operate at the international, the trans-governmental, and the transnational levels. The first value of this international approach is that it helps states and other actors to produce and share information more effectively, which increases the comparability of information and assessments from varied sources. The second value is its usefulness in coping with the problem of inconsistent or inappropriate national responses. Abbott provides a detailed elaboration of the framework approach to the "wicked" problem of innovation by drawing upon the structure and history of various international framework approaches as applied in other contexts.

Ruth Carter and Gary Marchant consider the strengths and weaknesses of principles based regulation, rather than rules-based regulation, as a means to address the complexity and rapid pace of innovation. Principles-based regulation tries to focus on desired outcomes rather than rigid rules by promulgating guiding principles that are broad, general and abstract. Regulated companies are intended to have considerable discretion about how to apply the principles to new situations. Carter and Marchant compare the principles-based and rules-based approaches, using the finance industry, which uses principles-based regulations, to draw distinctions. Principles-based regulations place the spirit of the law before the letter of the law; provide flexibility and freedom to regulated companies; better respond to the changing practices of evolving industries at lower cost; and can foster better relationships between regulated companies and the regulators. There are limitations to the principles-based approach: flexible principles can beget uncertainty; principles can ossify over time to resemble and function like rules; the transformation from rules-based regulations to principles-based regulations can be costly and time consuming; there can be substantial compliance problems because the regulator requires more information and cooperation from regulated companies to provide effective regulatory oversight; and a rules-based approach requires a change in the culture of the regulated companies and its industry to adopt a cooperative perspective on the role of the regulator in monitoring regulatory compliance. A special problem for a principles-based approach is technologies that are subject to regulatory oversight by more than one agency; it is unlikely that a principles-based approach can be implemented unless all involved agencies are pursuing principles-based regulation.

Lyn Gaudet and Gary Marchant explore four administrative law strategies that might be applied in a modified form to address issues raised by emerging technologies. The four tools are negotiated rulemaking, direct final rulemaking, online forums, and temporary legislation/sunset provisions. The chapter provides a brief introduction to administrative law and then addresses the four tools individually, elaborating upon their particular features and offering examples. Negotiated rulemaking attempts to join affected parties together in order to forge a consensus version of the proposed rule. This effort aims to avoid the delays and inefficiencies of an agency drafted rule that is contested in a lengthy notice and comment process followed by judicial review, perhaps in several iterations. Negotiated rulemaking is limited to situations where a limited number of interests are affected by the rule, a balanced representation of persons affected by the rule can be convened, there is a reasonable likelihood that consensus is achievable and the process will not cause, but rather avoid delay in the proposed rulemaking and the issuance of the rule.

Direct final rulemaking bypasses the elaborate notice and comment phase of most rulemaking. Instead, an agency drafts its proposed rule, provides public notice of the rule, and adopts it shortly thereafter without public input or comment. The motivation for this direct final approach is streamlined rulemaking, i.e., getting rules into effect promptly. Rules adopted in this manner must be uncontroversial because receipt by the proposing agency of any adverse comment or a notice on an intent to file an adverse comment leads directly to the withdrawal of the rule. Direct final rulemaking is well suited to minor changes to rules, e.g., cleaning up language ambiguities in a rule, but it problematic for proposed rules with encompassing or significant impacts.

Online rulemaking is the use of digital technologies to develop and implement regulations. It promotes access to a large quantity of information from a large number and variety of sources and increases public access and participation in the rulemaking process. Temporary legislation/sunset provisions involve the expiration of laws/rules after a specified period of time. This can be especially applicable to emerging technologies because regulations promulgated with limited information/significant uncertainties about such technologies must be revisited when they expire and more information/less uncertainty will lead to better phrased regulations at the subsequent revision. The regulations are likely more malleable and responsive when they must be redrafted periodically. Unfortunately, regulations that sunset are often reenacted without a careful reconsideration of their strengths and weaknesses.

Kathleen Waugh and Gary Marchant consider the possible use of collaborative voluntary programs to remedy part of the "pacing problem." They use environmental law as a lens to evaluate such programs because environmental law has a 20 year history of addressing the outputs of technology. Environmental law has pioneered regulatory experiments that attempt to avoid the shortcomings of command and control approaches, e.g., Best Available Technology (BAT), that lock in the status quo and provide no incentive for investment in new processes or technologies. Collaborative voluntary programs take various forms, but they strive to include regulated entities in regulatory target setting and emphasize flexibility about how those targets will be met. These collaborative programs often create institutional frameworks for on-going negotiations and innovation. Some regulatory models allow participating entities to essentially self-regulate. In general, leadership arises from multiple sources; solutions are framed to fit specific circumstances (instead of a one size fits all approach); potential for continuous improvements (rather than minimum compliance levels) is emphasized; and flexibility about how a regulated entity meets performance standards is a central feature.

The chapter evaluates the advantages and disadvantages of the collaborative voluntary approach for possible application to emerging technologies and provides a detailed description of several case studies to assist in that assessment. The five case studies offer several different kinds of collaborative voluntary programs with different methods of phrasing the relationship between the government regulators and the regulated entities. Two key themes are identified in the case studies: the value of information in generating better policies and strategies and the importance of flexibility for effective cooperation. Common problem areas were a widespread perception among environmental activists that undervalued cooperation as mere "cosmetics" and "greenwashing"; complaints by non-participating entities that participating entities were unfairly advantaged by the agreements; a concern that multi-stakeholder negotiations occurred on an inherently uneven playing field where information asymmetries benefited entities that were not candid about their operations and practices; and reservations about the proper role of government in regulating business and its objectionable entanglement in the complexities of an individual company's particular circumstances. Waugh and Marchant close with recommendations about future collaborative voluntary programs: flexibility is critical; rewards must offset the assumed risk; regulatory penalties are effective motivators; organization and industry dynamics are important selection criteria for program design; goals should be achievable, well-defined, and quantifiable; an effective multi-stakeholder process requires a level playing field; and the collaboration process needs to be adequately funded.

In summary, this volume endeavors to perform three tasks. First, it argues that the "pacing problem" is real and substantial. It cannot be dismissed as a largely mythical construct of over-active imaginations. The proof is amply demonstrated with quantifiable evidence and persuasive narratives. The portents of the problem are unexaggerated. Second, it explores the nature of this "pacing problem" by analyzing its impacts in several particular contexts. It poses hard questions for traditional notions of ethics and participatory democracy that cannot be ignored. Third, it provides several methods for possibly coping with the problem. These methods are suggestions; they are not meant to be either definitive or exhaustive. Indeed, one of the themes of this volume is the need for adaptive and flexible responses to the "pacing problem" that are amenable to modification and recalibration. A salient feature of emerging technologies is their rapid and non-linear development. It is unlikely that any single innovation in legal practice or ethical reflection will be able to encompass all of those developments. The methods described in this volume intend to provoke further consideration of other methods and approaches.

References

Buchanan, James M., and Tullock, Gordon. 1962. *The calculus of consent*. Ann Arbor, MI: University of Michigan Press.

Calabresi, Guido. 1982. *A common law for the age of statutes*. Cambridge, MA: Harvard University Press.
European Science Foundation. 2010. Human stem cell research and regenerative medicine: A European perspective on scientific. *Ethical and Legal Issues,* May 2010.
Fuller, Lon. 1964. *The morality of law*. New Haven, CT: Yale University Press.
Goldberg, Steven. 1994. *Culture clash: Law and science in America*. New York, NY: New York University.
Greenawalt, Kent. 1989. *Conflicts of law and morality*. New York: Oxford University Press.
Jasanoff, Sheila. 1990. *The fifth branch: Science advisers as policymakers*. Cambridge, MA: Harvard Univeristy Press.
Raz, Joseph. 1978. *The authority of law*. Oxford: Clarendon Press.
Snow, C.P. 1959. *Two cultures and the scientific revolution*. Cambridge: Cambridge University Press.

Contents

Contributors

Kenneth W. Abbott Arizona State University, Tempe, AZ, USA, ken.abbott@asu.edu

Braden R. Allenby Arizona State University, Tempe, AZ, USA, braden.allenby@asu.edu

Andrew Askland Arizona State University, Tempe, AZ, USA, sandy.askland@asu.edu

Ruth B. Carter Arizona State University, Tempe, AZ, USA, ruthbcarter@gmail.com

Lyn M. Gaudet Arizona State University, Tempe, AZ, USA, lyn.gaudet@gmail.com

Joseph R. Herkert Arizona State University, Tempe, AZ, USA, joseph.herkert@asu.edu

Deborah G. Johnson University of Virginia, Charlottesville, VA, USA, dgj7p@virginia.edu

Gary E. Marchant Arizona State University, Tempe, AZ, USA, gary.marchant@asu.edu

Lyria Bennett Moses University of New South Wales, Sydney, NSW, Australia, lyria@unsw.edu.au

Brian Rappert University of Exeter, Exeter, Devon, UK, B.Rappert@exeter.ac.uk

David Rejeski Woodrow Wilson International Center for Scholars, Washington, DC, USA, David.Rejeski@wilsoncenter.org

Daniel Sarewitz Arizona State University, Tempe, AZ, USA, daniel.sarewitz@asu.edu

Kathleen Waugh Arizona State University, Tempe, AZ, USA, Kathleen.Waugh@asu.edu

Part I
The "Pacing Problem"

Chapter 1
Governance and Technology Systems: The Challenge of Emerging Technologies

Braden R. Allenby

> *So long as we do not, through thinking, experience what is, we can never belong to what will be. . . . The flight into tradition, out of a combination of humility and presumption, can bring about nothing in itself other than self deception and blindness in relation to the historical moment.*
> *– Heidegger (1977)*[1]

1.1 Introduction: The Power of Technology Systems

It was 1804, in England. Matthew Murray invented what is usually considered the first railroad engine; later that same year, Richard Trevithick built a 40 pounds per square inch (psi) steam locomotive for the Welsh Penydarran Railroad. And the world changed forever.

It took a while. It was 1812 before the first commercially successful steam locomotives were put into operation on the Middleton Railway, and only in 1826 that the first rail line was laid in the United States, in Quincy, Massachusetts; only three miles in length, it was technologically obsolete in that the cars were pulled by horses. Optimistically, the Baltimore and Ohio Railroad was chartered only a year later, at first using sail power, then horse power – literally, as horses on treadmills were used to drive the carriage wheels. But by 1832, the *Atlantic*, weighing 6.5 tons and burning a ton of coal per round trip, hauled 50 tons from Baltimore on the B&O line over a distance of 40 miles at 12–15 mph (PSRM 2008; Schivelbusch 1977).

By the 1840s, it was clear that railroad technology was improving rapidly, especially in the United States, where long distance transportation systems were critical. Moreover, such technology was beginning, especially in a country like the United

---

[1]Heidegger, M. 1977. *The Question Concerning Technology and Other Essays.* (trans: Lovitt, W). New York, NY: Harper Torchbooks "The Turning," 49; "The Age of the World Picture," 136.

B.R. Allenby (✉)
Arizona State University, Tempe, AZ, USA
e-mail: braden.allenby@asu.edu

G.E. Marchant et al. (eds.), *The Growing Gap Between Emerging Technologies and Legal-Ethical Oversight*, The International Library of Ethics, Law and Technology 7,
DOI 10.1007/978-94-007-1356-7_1,

States, to have important cultural implications. Reflecting in part the growing realization of the national security and military implications of railroad technology and the advantage it gave the Union forces, President Abraham Lincoln in 1862 signed the Pacific Railway Act authorizing construction of the first transcontinental railroad. Seven years later, in 1869, two railroads and two corporations, the Central Pacific and the Union Pacific, met at Promontory Point, Utah, in a physical validation of Manifest Destiny (Nye 2003). But the United States was not, perhaps, the most dramatic demonstration of the integrated strategic and cultural power of railroad technology. Consider, for example, the rise of Prussia.[2] After the 1815 Congress of Vienna concluding the Napoleonic Wars, Prussia was simply another minor state, only one among many that littered Central Europe. In fact, it was more pathetic than most, because it was split into two pieces geographically, and those pieces were floating in the midst of strong and powerful states: Hanover, Hesse-Kassel, and, worse yet, France, Russia, Austria. But the brilliant Prussian military genius, Helmuth von Moltke, at least understood the power of railroads. He used them effectively when Prussia crushed the 1848 uprisings by using its railroads to rush troops from city to city, moving their military strength quickly to where it was needed. The Prussian military was impressed. In short order Prussia established the Prussian Railway Fund, a special fund that supported the construction of militarily critical lines that were not commercially viable. Dual use rail technology became de rigor; Prussian commercial freight cars were designed so that they could carry soldiers, horses, and military equipment if necessary. Unlike any other European power, the Prussians tied mobilization to rail networks; each Prussian regiment was assigned to a specific railhead where it would assemble when mobilized. Of course, innovation tends not to be an isolated event; the Prussians also had the most advanced rifle in Europe, the needle gun, world class military management, and highly advanced training. It all paid off in 1866 at the battle of Koniggratz, when the Prussians stunned – in fact, destroyed – a major European power, the Austrian Empire, in large part because they managed to transport 197,000 men and 55,000 horses to the front using railroads, a feat the Austrians had not imagined possible (Boot 2007). As Austria fell, Prussia rose; Koniggratz marked Prussia as a European power, even though it lacked the economy, the population, or the geographical advantages of others, such as France or England.

But the story doesn't end there, because, in large part as a result of their initial success using the railroad for military advantage, the Prussians, and thence the Germans, subsequently developed a strategic plan for continental warfare. The "Schlieffen Plan" was intended to enable victory by enabling the Germans to fight on two fronts by achieving a rapid strategic success on one front (against the French), then using the railroad to transfer their troops to the second front while their erstwhile opponent (Russia) was still mobilizing. This was, indeed, the German plan in World War I – in order to avoid a war on two fronts, which the Germans

[2]This example is taken from Allenby and Sarewitz (2011), and draws heavily on the description of this battle in Boot (2007).

recognized they probably couldn't win, quickly defeat the French and then rush those troops by rail to face the Russians, who everyone, including the Germans, knew would take a long time to mobilize. It almost worked. Only the unexpectedly rigorous French defense in the Battle of the Marne, and the fact that the Russians mobilized far more quickly than anyone anticipated, led to the Plan's failure and the stalemate of World War I trench warfare. So was it the lure of railroad technology that led the Germans to a fatal mistake, and perhaps encouraged them to begin World War I? Without Konniggratz, and the reliance which verged on faith on railroad technology, would Germany have initiated the hostilities that led to two World Wars and the end of the naïve nineteenth century faith in Enlightenment progress and rationality?

To moderns, railroad technology is familiar, even trite; to the people that lived during the era when railroad technology diffused across the global landscape, it brought enormous change. It was an inexorable juggernaut of change: not just technologies, and family businesses, and small farms, and transport change, but deep, unsettling cultural and psychological change. The railroad built a new world, but not before destroying much of the old, the comfortable, the familiar.

For many people, for example, there is nothing more familiar than time – if they even think about it at all. Yet modern time is a product of railroad technology, for as a regional and national scale integrated transportation network, railroads required a uniform, precise system of time that was coextensive with the network – they required "industrial time" (Rosenberg and Birdzell 1986). Before railroads, local times were isolated and charmingly idiosyncratic: London time was four minutes ahead of Reading, for example, and fourteen minutes ahead of Bridgewater (Schivelbusch 1977). In fact, the industrialization of time can be seen as a co-evolutionary process as railroad technology matures: for a considerable time in the US, each train company had its own time, so that Buffalo at one time has three different clocks, Pittsburgh six, reflecting the number of railroad companies using their stations. Regional standard time did not gain legal recognition in the US until 1918 (Schivelbusch 1997). Large integrated networks also require co-extensive communications systems if they are to be coordinated, so the railroads called forth the telegraph (Grubler 1998).

Just as they devoured miles, railroads devoured the simpler, localized economic institutions that characterized a largely agrarian America (a pattern less obvious in Europe). Railroad firms needed far more capital than the simpler factory capitalism they replaced; they thus required, and shaped, modern capital and financial markets (railroad construction was the single most important stimulus to industrial growth in Western Europe by 1840s) (Freeman and Louca 2001). Similarly, railroads fundamentally changed economic and power structures; especially in the US, railroads restructured the economy from local and small regional business concentrations to trusts (scale economies of national markets). Big Sugar, the Tobacco Trust, Standard Oil ... industrial monopolies of unprecedented scale rode in on rails of steel. With the railroad, economic power passed to industrial firms from agriculture; more subtly, so did cultural authority (Marx 1964; Nye 1994). The transformative touch of the railroad not only created national institutions, but re-structured the continent itself, writing a new biological chapter, a new demographic chapter. Railroads transformed landscapes at all scales, both physically and psychologically. Chicago existed, and structured the Midwest economically, politically, physically and environmentally, because of railroads (Cronon 1991). Psychologically, railroad technology did not just extend, but obliterated, the sense of place and rhythm that previous transportation technologies, the horse and carriage and canal, had encouraged. Schivelbusch (1986 pp. xiii, xiv) notes that railroad technology "introduced a new system of behavior: not only of travel and communication but of throught, of feeling, of expectation. travel by railroad inevitably (if unconsciously) assimilated the personal traveler into a physical system for moving goods." The space-time compression so beloved of post-modernists is not a purely modern phenomenon, for it is a common complaint of rail travelers in their day. Not for nothing did the Impressionists celebrate rail stations as images of modernity and urban splendor (with perhaps a touch of occasional alienation).[3]

Perhaps most fundamentally, however, in the United States particularly railroads became a potent symbol of national power, and, more subtly, instantiated and validated the American integration of religion, morality and technology (Marx 1964; Nye 1994). In doing so, the railroads fundamentally and radically shifted the dominant American worldview from Jeffersonian agrarianism to a high technology, capitalist sublime, from an Edenic to a high-tech New Jerusalem teleology (Marx 1964; Nye 1994). The "shining city on the hill" was not to be an equitable democracy of farmers, but a technological overcoming of human frailty. This is a powerful and potentious change that even today replays itself as the sustainability and environmental discourses yearn for the Edenic and the industrial, commercial, and science and technology communities strive to build their New Jerusalem (Allenby 2005). It also flavors dialogs between the United States and the European Union, particularly around agricultural policy (and genetically modified organisms) which for the US is an industrial issue, and for the E.U. much more of an agrarian landscape issue.

Railroads are only one example of what economic historians call "long waves" of innovation (sometimes called "Kondratiev waves"), where technology clusters develop around core technologies (much as the telegraph did with the railroad),

[3]While the best known series of Impressionist paintings of the railroad is probably Monet's seven paintings of the Gare Saint-Lazare in Paris, which offer a wonderful treatment of the new technology, Manet's *The Railway*, also known as the *Gare Saint-Lazare*, is more nuanced, with the characters in the foreground separated from the railroad by an iron fence, and the adult turning away from the steam of the railroad even as the child peers through the bars. Nonetheless, contemporary observers could not contain their passion for the new technology: as Jacques de Bies exclaimed in his lecture "Edouard Manet" in the Salles des Capucines, in Paris on January 22, 1884 (quoted in National Gallery of Art, 1998),

> It's true, the locomotive is missing and one does not see the train. The smoke is enough for me, because it denotes the fire, which is like the soul of the engine. And the engine, as you who are listening know well, is the intelligence, the glory, and the fortune of our century. For future generations, our nineteenth century will be a locomotive, just as papal Rome is a tiara, as Venice is a gondola, . . . and as our French Middle Ages is the armor of a baron.

with accompanying social, cultural, institutional, policy, legal and economic change. The mechanization of textile manufacture was the basis of the Industrial Revolution in the UK; railroads and steam technology powered a wave from about 1840 to 1890; steel, heavy engineering and electricity, from about 1890 to 1930; the automobile, petroleum, and aircraft from about 1930 to 1990; the information cluster with its computerization of the economy, from about 1990 to the present (Freeman and Louca 2001). Each of these technology clusters co-evolves with profound and unpredictable institutional, organizational, economic, cultural and political change. Thus, for example, prior to the railroads the American economic system was characterized by very loosely coupled local economies; after the railroad came the trusts and national monopolies. From the perspective of the former, it would have been hard indeed to predict the development of these new economic institutions – much less the legal responses that eventually were required as economic governance itself shifted to meet new conditions (Bruchey 1980) . Similarly, specialized professional managerial systems and associated "Taylorism" industrial efficiency techniques characterized the heavy industry cluster, while a far more networked, flexible structure began to evolve during the information cluster (Castells 2000). Each of these technology constellations carries with it unpredictable social and institutional implications; each requires new policies and new legal structures.

The railroad example does, however, make clear several important principles of technological evolution. For one thing, technologies must be understood as integrated cultural, economic, institutional, and built phenomena. To consider technology to be merely artifacts, while appropriate in many cases, would lead to gross over-simplification and dysfunctional analysis and policy formulation if applied to technology systems (Bijker et al. 1997). Secondly, any technology of any significance tends to be profoundly destabilizing of existing norms, institutions, employment patterns, firms, and power relationships – and thus, of course, to the degree it is so, it will generate substantial and potentially powerful opposition. Whether a culture will evolve technologically thus depends to a not insignificant extent on how powerful its conservative forces are, and whether they are able to merely hinder technological evolution (as in the case of stem cells and the Bush Administration), or to stifle it completely (as China did with its technology in the early 1400s, when it was by most estimates the most technologically advanced country on earth) (Landes 1998). Thirdly, because of the complexity of technology systems, especially as they are manifested in cultural, institutional, and social domains, projecting the effects of technology systems before they are actually adopted is not just hard but probably impossible.

These observations may seem trivial given the railroad example, but that is only because we're now in the position of looking back, when the traumas, conflicts, and dislocations are safely played out, and for the most part forgotten. We are, in a meaningful sense, children of the railroad, and we could no more conceive of a world that had not been shaped by that technology than we could imagine living on the moon. Moreover, if the history of technological evolution is a warning, it is an inadequate one for the wave bearing down on us now. Technological change is always potent, but now we have not just one or two enabling technologies undergoing rapid evolution, we have five: nanotechnology, biotechnology, robotics, information and communication technology (ICT), and applied cognitive science (NBRIC).

1.2 The Five Horsemen of Emerging Technologies

Taken as a whole, the NBRIC technologies in some ways are the logical end of the chapter of human history that began 2,500 years ago with the Greeks. Nanotechnology is the culmination of material science and chemistry, in that it extends human will and design to the atomic level. As for biotechnology, J. R. McNeill, an environmental historian, notes that (2000 at 193–94):

> By the twentieth century, our numbers, our high-energy technologies, and our refined division of labor with its exchange economy made us capable of total transformation of any and all ecosystems. ... In the twentieth century we became what most cultures long imagined us to be: lords of the biosphere.

Robotics continues to expand its exotic menagerie; spook agencies are developing robotic insects that can be used for surveillance, while hybrid robots – hardware robots guided by hybrid rat neuron/chip configurations – undergo testing in the laboratory (in the instant case, some 300,000 rat neurons in a soup – the image is of the proverbial brain in a vat, only it's real and running a robot (Marks 2008)). Military use of robots is expanding greatly, from weapon platforms that are manipulated at a distance to jet fighters that are designed so they can either be flown directly by a pilot – or from a distance. More subtly, different cultures envision very different trajectories as cyborg states proliferate (consider the modern soldier, with all his or her instrumentation, GPS capability, and the like). Americans and Europeans seem to envision a somewhat dystopian relationship, with technology in conflict with the human; Japanese, Korean, and Chinese cultures seem to envision an integration of the human and the technological. Thus, Americans tend to view robots that are designed to look and behave like humans as somewhat spooky; Japanese like them. In popular media, the Western Frankenstein perspective shows up in movies like *The Matrix,* whereas the Asian perspective is quite common in anime, such as the two *Ghost in the Shell* movies. It is interesting but entirely speculative to suggest that this cultural difference may in the long run prove to be a source of significant cultural advantage for Asian countries.

ICT gives us the ability to create virtual worlds at will, and facilitates a migration of functionality to information rather than physical structures. Thus, money used to be coins and paper bills, themselves mere symbols of value, but now even that physical premise is gone. Money is electrons somewhere in cyberspace, and financial instruments have become so mathematical that no one can figure out anymore which shell the risk is hidden under. But ICT has more subtle implications. Consider, for example, how Google™ enables real time recovery of virtual any fact on the Internet – or, put another way, how memory as a cognitive function has now been distributed across the Web, just as in earlier eras poetry and religion, upon being printed, became available to everyone, not just elites. Moreover, the acceleration of change and accumulation of information also accelerates changes in the structures – cultural constructs, ideas and ideologies, even language itself – that humans use to simplify and symbolize their world, so that assumptions about meaning can no longer be regarded as stable.

Finally, there's cognitive science, which is particularly challenging in that it affects not just external systems, but the human itself. In other words, where other technology systems made fundamental changes, at least they were external to the core concept and physicality of human, whereas cognitive science accelerates the process of modifying (or designing, or actively intervening in) the human itself. This is difficult ground – what is "the human" to begin with? – but at least in some cases the general idea behind the technology is familiar, even if the technology is new. Thus, for example, humans have been using drugs, from alcohol and caffeine to more esoteric (and less legal) formulations, for thousands of years, but the new cognitive enhancers – Ritalin for concentration, modafinil (Provigil) for alertness, and many others in the pharmaceutical pipeline – are not only coming more quickly but are increasingly being designed for specific cognitive effects. Similarly, the printing press meant that individuals no longer needed to memorize Homer, but could go to the page or chapter they wanted – which was a form of diffusion of memory. In that sense, Google™ is a continuation; in the power of integrated external memory, however, Google™ and similar systems are no less transformative than printing. Lest we focus too much on the individual level, however, it is worth noting that without the printing press there could not have been the Protestant Reformation (because individuals until then did not have access to written material in general, and the Bible in particular, so a theology of direct relation to God through scripture could not have arisen); and the Reformation launched three centuries of bloody warfare across Europe. Cognitive technologies are neither neutral nor gentle.

To take another example, cognitive research has been for a number of years now quietly but effectively undermining the concept of free will by, among other things, increasingly demonstrating that consciousness appears to be an afterthought (literally) rather than an agent of action (Baer et al. 2008). From another angle, geneticists continue to demonstrate that certain genetic patterns are associated with a high probability with complex behavior patterns including violence and antisocial behaviors. Clearly, such work raises difficult issues for the legal system, both metaphysical (e.g., the dependence of criminal law on free will and conscious intent), and substantive (what level of pre-criminal activity restraint based on, e.g., genetic propensities, is permissible? How is protection of society balanced against the rights of the individual when we gain power to rank individuals against probabilities of different types of antisocial behaviors?).

Before one can begin to answer such questions, however, one must perceive the relevant context. This is increasingly problematic: the public (and, indeed, technologists outside of their particular specialties) get only impressionistic glimpses of emerging technologies in stories about particular events or experimental results. Virtually no organization, and certainly no individuals, perceive what's actually happening across the technological frontier, or even in particular domains. It is therefore not at all surprising that most people are completely blind to how real cognitive science scenarios that they might regard as science fiction actually are. Thus, for example, the US Army recently funded researchers at the University of California at Irvine, Carnegie Mellon University, and the University of Maryland to study how to design and build "thought helmets" for soldiers that detect and interpret brain waves in terms of individual words, and rebroadcast them as radiowaves to other soldiers – in other words, telepathy (Physorg 2008). While application remains years away, this builds on much other work that's already been done at places like Carnegie Mellon University, that indicates that interpretation of individual words from brain activity is doable today. Combine this with the Duke University monkey experiment where a monkey learned to manipulate a remote arm via brain waves and wireless transmission (and subsequently to make a robot in Japan run), and the breakout of the brain from the Cartesian model of the individual human is virtually complete (or, looked at another way, the implicit boundary between the individual human and environment is breached beyond redemption). Already, for example, it is possible to purchase, for a list price of under $300, a headdress device that reads one's brainwaves and enables direct mental control of avatars in synthetic realities. Primitive at this stage? Absolutely. But the implications are obvious – and, importantly, these technologies are being driven by national security and defense needs, and powerful financial incentives (technologies that make games more attractive can be worth billions). Thus, even in this limited domain a constellation of recent results integrates into a culturally and psychologically powerful substantive change in cognitive and perceptual systems, with huge legal and policy implications in domains from privacy to anti-terrorism, systems security, and military and national security.

But it is not just that each NBRIC technology system is powerful; it is that they are combining in unexpected ways that are both beyond any single technological domain, and very potent. The combined rat neuron/robot is an example; the genetic/cognitive science collaboration on behaviors another; an ICT-based helmet that reads thoughts yet another. Another that will no doubt raise serious policy and legal issues going forward as society struggles to adjust in response is radical human life extension.

At one extreme, some predict the achievement of "functional human immortality" within 50 years, either as a result of continuing advances in biotechnology, or as ICT and computational power enable downloading of human consciousness into information networks (Moravec 1988; Kurzweil 2005). While such predictions are viewed by most experts as highly unlikely, there is a growing consensus that substantial extensions of average lifespan, with a high quality of life, are achievable in the next few decades (De Grey 2004). For example, the IEEE *Spectrum,* a mainstream technical journal, ran a series of articles in 2004 on engineering and aging which concluded that using "engineered negligible senescence" to control aging will allow average ages of well over 100 within a few decades. Others, while not rejecting the possibility, are more skeptical (Vijg and Campisi 2008). Historically, substantial increases in lifespan are not unprecedented; before 1800, life expectancy virtually everywhere was only 30–35 years; South Korea has seen a 30 year jump in lifespan since 1960 (Clark 2007; The Economist 2008). Moreover, what makes radical lifespan extension interesting is not just the integration of technological domains – complexity and network theory, genetics, biotechnology, electrical engineering theory and methodology, and much more – but how unaware virtually all policymakers and the public generally are of the possibility. This is even more bemusing because of the obviously challenging implications (for pension and old-age systems, reproductive freedom and demographic trends, and material and energy consumption patterns, for example). To add to the conceptual challenges, radical lifespan extension is only a small, albeit emotionally potent, area of human enhancement research; taken together, the suggestion that "the human" is in the process of becoming a new arena of systemic design is not implausible, although of course specifics are unpredictable. N. Katherine Hayles, for example, in her aptly named book, *How We Became Posthuman,* traces the evolution of the posthuman through the concepts of homeostasis, then reflexivity, then, finally, virtuality. While Hayles is cautious about the implications of this on-going and accelerating process, some foresee enormous potential: Roco and Bainbridge in an NSF report entitled *Converging Technologies for Improving Human Performance*, for example, conclude (2003 p. ix):

> With proper attention to ethical issues and societal needs, converging technologies could achieve a tremendous improvement in human abilities, societal outcomes, the nations's productivity, and the quality of life. ... Examples of payoffs may include improving work efficiency and learning, enhancing individual sensory and cognitive capabilities, revolutionary changes in healthcare, improving both individual and group creativity, highly effective communication techniques including brain-to-brain interaction, perfecting human—machine interfaces including neuromorphic engineering, sustainable and "intelligent" environments including neuro-ergonomics, enhancing human capabilities for defense purposes, reaching sustainable development using NBIC tools, and ameliorating the physical and cognitive decline that is common to the aging mind.

1.3 Technology, Complexity and Earth Systems Engineering and Management

The evolution of new core technologies – actually, a suite of them in the form of NBRIC – dramatically affects our usual assumptions about stability and cultural frameworks along three dimensions: complexity, contingency, and accelerating change. This can be briefly if simplistically captured in the understanding that accelerating technological evolution has decoupled from social, cultural, and institutional evolution, and that as a result of that decoupling, we are increasingly incapable of managing the complexity that we have ourselves created, and unable to deal with the level of contingency in foundational beliefs and perceptions that technological evolution increasingly generates. This sounds somewhat ethereal, but the implications of failing to appreciate the power of complexity can be quite important. An obvious example is the end of the Cold War, when primitive communism in the Soviet Union and China collapsed not from external conquest, but because the centralized economic model adopted by large Marxist societies simply became incapable of managing the complexity inherent in a modern industrial economy. Simply put, the

Soviet model assumed a knowable, simple system whereas the decentralized information structure of the market model adopted by the West proved far more adept at managing complexity.

To begin with, we face radically increasing complexity of at least four different kinds. The most intuitive is static complexity: the systems within which we must function contain increasing numbers of components, subsystems, and linkages among them. Supply chains become supply networks as even relatively simple products instantiate global networks; medicine becomes molecular; socially, communities of all types spring up on the Internet. The next stage is, of course, dynamic complexity, as these networks evolve over time. The two are often related, but do not need to be; the famous beer game that confounds MBA students as they try to balance a simple supply chain consists of only a wholesaler, a brewer, and a retailer (the complexity arises with the lag times introduced between the demand and supply functions of these three nodes) (Senge 1990). This leads to the third kind of complexity, which arises as technological or environmental systems become entangled with human systems, "wicked complexity." Wicked complexity is difficult precisely because it entrains consciousness, intentionality, reflexivity, and other characteristics of human behavior at individual and institutional scales. A study of a salt marsh, for example, does not change the salt marsh, as the information domains (the study or report, and the physical and biological systems of the salt marsh) remain separate; a study of New York City, however, reflexively becomes part of the information structure of the City, and thus changes the dynamics of the system in ways that the study could not predict. Wicked systems are thus inherently uncertain and unpredictable, and require real time interaction, because until behavior actually emerges, future system states are uncertain.

The final level of complexity is scale, which arises because we must begin to design, engineer and manage integrated human/natural/built earth systems at not just national, but regional and global scale. We live in a world that is fundamentally different from anything that we have known in the past, and a major reason is the technology systems that, in NBRIC, are approaching rates of autocatalysis and change that are unprecedented, and that mark a demarcation not just between pre-modern and modern, but between humans as a "natural" species, and humans as a "technological" species. In one sense, of course, we have simply begun to perceive that which thousands of years of human history have created, although the Industrial Revolution undeniably accelerated the process – a world dominated by one species and the activities and products characteristic of that species, from automobiles to cities to the creation of vast new cyberspaces. It is a world where the critical dynamics of major earth systems, be they atmospheric, biological or radiative, or for that matter cultural, economic, or technological, increasingly bear the imprint of the human. Several examples should suffice to provide a flavor of this anthropogenic Earth. Consider that most physical aspect of a body in space, its radiation spectrum, and then recall that the Earth's spectrum is no longer just a matter of reflections from clouds, emitted infrared radiation, and the like, as it has been for billions of years until now. Rather, it includes television and radio broadcasts, and leakage from all sorts of technologies – a reality captured brilliantly in that picture of the Earth from space at night, with the electric lights spread over North America, Europe and Asia. Today, perhaps the most fundamental physical aspect of our planet, its radiation spectrum, carries our signature on it – and, underlying that profound change is the core technology of another "long wave," electricity.

Another obvious example is global climate change, which, despite the limited perspective of the current negotiation process, represents a new reality: fitful and ad hoc as our current responses are, they stand for the dawning of a realization that, regardless of short term responses such as the Kyoto Treaty, our species will be engaged in a dialog with our climate, our atmospheric chemistry and physics, and the carbon cycle so long as we exist at anywhere near our current numbers on the planet. We can reduce – more likely, redistribute – some of our impacts on these complicated and interrelated systems – biofuels, for example, attempt to address carbon cycle management, but at the expense of nitrogen and phosphorous cycles at this point – but we will not eliminate the growing human influence. Moreover, these particular perturbations are all part of interconnected global systems, and a population of over six billion humans, each seeking a better life, ensures that our overall role in global systems will increase absent some sort of population crash. And underneath this phenomenon is another "long wave" constellation of technologies, including the fossil fuel energy economy and the automobile.

A third example might be the restructuring of the biosphere that is currently beginning. This is more complex than the usual framing of it as a "crisis in biodiversity," which is a partial perception, because even if the decrease in evolved biodiversity is as steep as alleged, it only involves a part of biology – "evolved biology." It thus overlooks the project of understanding and designing new forms of life, a series of efforts, from genetics to agricultural science, that are coalescing into a new field called "synthetic biology." Synthetic biology merges engineering with biology by, among other things, creating standard biological components that can be mixed and matched in organisms to provide desired functions. This allows researchers to treat biological pathways as if they were components or circuits, and to create organisms from scratch – not to mention extending beyond existing biological systems by, for example, creating life based on different genetic codes than those found in the wild (MIT, for example, has established a Registry of Standard Biological Parts, called "BioBricks," that can be ordered and plugged into cells, just like electronic components). Others have, somewhat controversially, assembled a number of viruses from scratch, including the viruses for polio and the 1918 flu epidemic. Researchers have also engineered the genes of *Escherichia coli* to incorporate a 21st amino acid, opening up an option space for design of biological organisms that has been unavailable to evolved biological systems for billions of years. Commercialization of synthetic biology products continues to accelerate; although the most well known examples are agricultural genetically modified organisms (GMOs), the technology is far more widespread. According to the Economist, in 2004 some 5 percent of world chemical output was estimated to derive from genetically engineered technologies, and figures for biotechnology patent filings in OECD countries continue to rise sharply.

Synthetic biology does not just reconfigure the biological sciences; the potential implications are far more profound. Most obviously perhaps, the definitions and taxonomies that are inherent in traditional biology are destabilized by synthetic biology. The definition of a species, for example, as a group of organisms able to interbreed with one another and produce fertile offspring is increasingly irrelevant when biological systems are built with BioBricks. Moreover, biodiversity becomes a product of design choices, and industrial and political imperatives (security issues, for example), rather than evolutionary pressures. Evolutionary biology tends to create stable systems (because the components co-evolve), but synthetic biology tends to create efficient designs; this suggests that resilience and stability of bio-based systems might be less in a regime characterized by increasing replacement of evolved biology by designed biology.

More broadly, reconfiguring biological systems as design spaces, and commoditization of resulting biological products means that biological systems increasingly express the dynamics of human systems (wicked complexity), so that understanding biological systems increasingly requires an understanding of the relevant human systems, and the contingency that characterizes human systems comes to characterize biological systems. To take an obvious marketing example from conservation biology, in an arbitrary and profoundly cultural process some species are preserved because they are charismatic megafauna: pandas, tigers, or whales. Many, many others go extinct because they are only insects, or plants, or ugly, or unknown; a few, like smallpox, because humans detest and fear them (with the important proviso that, in an age of biotech, national security and terrorism, extinction, at least for viruses and bacteria, is never forever).

In short, biology is increasingly becoming a cultural science. But note that in order to even ask such questions and raise such concerns, it is necessary to perceive the importance of synthetic biology, a perspective that the current institutional framework of the biological disciplines tends to discourage.

The complexity introduced by emerging technological systems is augmented by their destabilizing effect on cultural constructs, ideologies, and institutional verities. Of course, such concepts and frameworks change over time, but if the time is long enough (seemingly a human generation or so), and they are captured in familiar forms such as laws, regulations, or social patterns and habits, they appear to be stable to the people living at the time. They can thus be assumed as forming a framework within which operationally and ethically responsible decisionmaking can occur. But as technological change accelerates, the destabilizing effects shorten, and become visible and, indeed, troubling to the societies affected. Thus, the current pace and depth of technological change has a twofold effect: it renders contingent much of what we consider to be fixed, and, because it turns the human into a design space, it also removes the stability of the one fixed point – ourselves as human beings – we have generally taken as a reference point. In any assessment, the question is usually what can be assumed a priori; we have few methodologies, and fewer legal and policy structures, which are competent when both the subject of analysis, and the assumptions upon which the analysis is based, are understood to be contingent.

Obviously, the continuing shift of the human from an a priori assumption to a design space substantially enhances the contingency of Earth systems of all sorts.

But technological change has other effects that enhance unpredictability and contingency, for complexity and emergent behaviors, though arising from technological systems, are not simply in the technological realm. To take an obvious example, it is difficult enough to understand the technological implications of the Internet: distribution of memory through Google™, increasing integration of synthetic and "real" realities, substitution of social for physical networking, security of increasingly coupled systems arising in many different domains. But the Net has profound cultural implications as well, for as it accelerates and augments information flows, it inevitably affects culture, which is, after all, a domain heavily influenced by information. Consider, for example, the impact of the Internet on what sociologists call "cultural constructs" in the context of environmental debates. In particular, many of the terms that characterize the environmentalist discourse, and that are taken as self-evidently "real" and thus foundational for purposes of policy, discourse, and ethics – "sustainability," "nature," "wilderness," or "environment," for example – are, in fact, highly contingent cultural constructs. That is, they reflect not eternal verities that are "out there" as much as they do particular ways of thinking about the world that reflect a specific cultural and historical context (Cronon 1991; Hacking 1999). As Merchant (1995 at153) observes, "Nature, wilderness and civilization are socially constructed concepts that change over time and serve as stage settings in the progressive narrative." "Sustainable development" is the most obvious example. Explicitly delineated and popularized in a specific book, *Our Common Future* (WCED 1987), the cultural construct of "sustainable development" and the looser term of "sustainability" have over the past 20 years become major goals of environmental and social policy, international negotiations, and even domestic and international law. Thus, what began as a quite obvious cultural construct now defines for many the desired endpoint for much human activity – the teleology of sustainability. In the process, the contingency of the term – although quite explicit – has become invisible to most people.

Why bring up cultural constructs at all? Because, while the uncertainty and contingency of these evolving systems precludes any real knowledge about how these constructs will be interpreted in future, what can be predicted with significant certainty is that they will be very different than today. Yet much of our current legislation and international negotiation processes assume their stability: we want to keep climate where it is today; we attempt to privilege current constructs of "wilderness," "nature," and "biological diversity" over periods not just of decades but of centuries even as the evolution of ICT shortens the average life of such constructs. From a technology perspective, to the extent one is designing traditional artifacts – a toaster, an automobile – cultural constructs will remain fixed. But the level of earth systems engineering and management – say, designing and supporting the continued evolution of the Everglades, or designing components of the carbon cycle to manage climate perturbations, or trying to understand and work with the evolution of NBRIC technologies – is a systems function, a continuing dialog, that extends well beyond the cycle time of many cultural constructs even now. More rapidly changing cultural constructs will lead to even more contingency and unpredictability in such engineering and management challenges, because the basis for policy and dialog – the cultural constructs that are common to a particular period and discourse –

will be shifting more rapidly. This may be one of the most important, and difficult to study and understand, implications of the evolution of ICT specifically and emerging technologies generally.

1.4 Conclusion

Even this brief discussion makes several important points regarding emerging technologies. Perhaps most importantly, technological change cannot be understood as an isolated event. Rather, it represents movements towards new, locally stable, earth systems states. These states integrate natural, environmental, cultural, theological, institutional, financial, managerial, technological, built and human dimensions; they change our worldviews, our cultural values, and even construct our sense of time. Technologies do not define these integrated earth system states, except by convenience, but they are an important destabilizing force across society. As such, they generate not just new opportunity but also continuing transition costs – as Joseph Schumpeter famously characterized it in *Capitalism, Socialism and Democracy* (1942), the "gale of creative destruction" that is innovative capitalism. They also thus generate occasionally strong opposition, especially from cultural groups that are either marginalized, or whose values are threatened by continuing technological evolution. From a legal perspective, the law becomes an important playing field upon which economic and institutional interests that are threatened try to strangle new technologies before they can be launched and locked in to political and economic systems.

In part, this is because technology is a major means by which humans have expressed their will to power. The implications of this observation are not just academic. First, many specific applications and developments across the technological frontiers discussed above will occur in the context of active conflict, such as the Battle of Koniggratz mentioned above – or the wars in Iraq and Afghanistan, where opposition to its deployment will be perceived as verging on treasonous. More subtly, cultures that develop technology, and encourage the development of frameworks within which it can react upon itself and so accelerate its own evolution, tend to thereby gain power over competitors. Because technologies create such powerful comparative advantages as between cultures, those cultures that attempt to block technology will, all things equal, eventually be dominated by those that embrace it. The implications of these dynamics, taken together, is that it is likely that technological evolution will be difficult, if not impossible, to stop. Whether and how it can be managed in the age of global cultural competition and globalized technological competence becomes an important research question.

It would be difficult enough to deal with the challenges of emerging technologies if they were evolving at historically traditional rates. The data indicate, however, that the rate of technological change is not slowing, but rather accelerating dramatically. This has a number of implications, some of which are sketched above. But in a globalizing culture, it will become an increasing source of dissatisfaction for more and more people, for it is stretching the bimodal distribution between those who constitute the global elite and who, primarily through education and culture, are able to prosper under such conditions, and those who are left behind. The latter have a strong tendency to seek stability in outmoded ideologies and fundamentalist movements – religious, environmental, social. These movements are desperate responses to a world that, for such individuals, has become irrational, and, as it destabilizes patterns of belief and behavior they invest with meaning, profoundly challenging and frequently evil.

Understanding these systems enough to try to legislate or regulate is also challenged because current technological evolution is arguably unprecedented. Previous technology clusters revolved around one or perhaps two evolving technologies – say, rails and steam, or automobiles and petroleum. The constellation of nanotechnology, biotechnology, robotics, ICT, and cognitive science, however, marks a culmination of sorts of traditional technological evolution, for among other things it extends control of materials to the atomic scale, and lays the groundwork for the complete integration of the human and the technological. The Earth, biology, and indeed even the human itself become design spaces and, in doing so, render contingent virtually all of what we have taken to be fixed.

And so we have the question that Lenin asked at the turn of the last century: What is to be done? First, the institutional ability to systematically perceive and understand the frontiers of technological evolution as integrated entities must be developed, for any action taken without such perception, while perhaps justified by local conditions, is spasm rather than sentience. Second, we need to understand the dynamics of these systems far better than we do, for only then can we begin to understand how we may best interact with them to achieve more desirable outcomes and trajectories. The operating assumption at this point – that we both understand these systems, and are capable of managing them so that we achieve desired outcomes without unfortunate unanticipated consequences – is at best whistling in the dark, and more likely an abdication of ethical and rational responsibility. And, finally, we need to develop institutional mechanisms, including legal and regulatory tools, which engage with these systems in ways that are both productive and predictable. While the theoretical mechanisms for doing so are being suggested (e.g., real time technology assessment, and real time macroethical assessment (Allenby and Sarewitz, 2011)), the operational capability is virtually nil at this point. We have our work cut out for us.

References

Allenby, B.R. 2005. *Reconstructing earth.* Washington, DC: Island Press.

Allenby, B.R., and D. Sarewitz. 2011. *The Techno-Human Condition.* Cambridge: MIT Press.

Baer, J., J.C. Kaufman, and R.F. Baumeister. 2008. *Are we free: Psychology and free will.* Oxford: Oxford University Press.

Bijker, W. E., T. P. Hughes, and T. Pinch, eds. 1997. *The social construction of technological systems.* Cambridge: MIT Press.

Boot, M. 2007. *War made new.* New York, NY: Gotham.

Bruchey, S.W. 1980. *Small business in American life.* New York, NY: Columbia University Press.

Castells, M. 2000. *The Rise of the Network Society.* Oxford: Blackwell Press.
Clark, G. 2007. *A fairwell to alms: A brief economic history of the world.* Princeton, NJ: Princeton University Press.
Cronon, W., ed. 1991. *Nature's Metropolis: Chicago and the Great West.* New York, NY: W. W. Norton and Company.
De Grey, A.D. N.J. 2004. *Strategies for engineered negligible senescence.* Annals of the New York Academy of Sciences 1019.
Freeman, C., and F. Louca. 2001. *As time goes by: From the industrial revolutions to the information revolution.* Oxford: Oxford University Press.
Grubler, A. 1998. *Technology and global change.* Cambridge: Cambridge University Press.
Hacking, I. 1999. *The social construction of what?* Cambridge: Harvard University Press.
Hayles, N.K. 1999. *How we became posthuman.* Chicago: University of Chicago Press.
Heidegger, M. 1977. *The question concerning technology and other essays* (trans: Lovitt, W.). New York, NY: Harper Torchbooks.
IEEE *Spectrum.* 2004. Engineering and aging. *Spectrum* 41 (9): 10, 31–35.
Kurzweil, R. 2005. *The singularity is near.* New York, NY: Viking.
Landes, D.S. 1998. *The wealth and poverty of nations.* New York, NY: W. W. Norton & Co.
Marks, P. 2008. Rat-brained robots take their first steps. *NewScientist* 199 (2669): 22–23.
Marx, L. 1964. *The machine in the garden: Technology and the pastoral ideal in America.* Oxford: Oxford University Press.
McNeill, J.R. 2000. *Something new under the sun.* New York, NY: W. W. Norton & Company.
Merchant, C. 1995. Reinventing Eden: Western culture as a recovery narrative. In *Uncommon ground: Rethinking the human place in nature*, ed. W. Cronon. New York, NY: W. W. Norton and Company.
Moravec, H. 1988. *Mind children: The future of robot and human intelligence.* Cambridge: Harvard University Press.
National Gallery of Art. 1998. *Manet, Monet, and the Gare Saint-Lazare.* Exhibition brochure. Washington, DC: National Gallery of Art.
Nye, D.E. 1994. *American technological sublime.* Cambridge: MIT Press.
Nye, D.E. 2003. *America as second creation: Technology and narratives of new beginning.* Cambridge: MIT Press.
Physorg. 2008. U.S. Army invests in 'thought helmet' technology for voiceless communication. Physorg.com, www.physorg.com/printnews.php?newsid=141314439, Accessed 24 Sept 2008.
PSRM (Pacific Southwest Railway Museum). 2008. *Railroad history: Important milestones in English and American Railway Development.* www.sdrm.org/history/timeline. Accessed Oct 2008.
Roco, M.C., and W.S. Bainbridge, eds. 2003. *Converging technologies for improving human performance.* Dordrecht: Kluwer Academic Publishers.
Rosenberg, N., and L.E. Birdzell, Jr. 1986. *How the west grew rich: The economic transformation of the industrial world.* New York, NY: Basic Books.
Schivelbusch, W. 1977. *The railway journey: The industrialization of time and space in the 19th century.* Berkeley, CA: University of California Press.
Schumpeter, J.A. 1942 (reissued 2008). *Capitalism, socialism and democracy.* New York, NY: Harper Perennial Modern Classics.
Senge, P.M. 1990. *The fifth discipline.* New York, NY: Doubleday.
The Economist. 2008. The odd couple: a special report on the Koreas (center Section, 27 September 2008). *The Economist.*
Vijg, J. and J. Campisi. 2008. *Puzzles, promises and a cure for aging. Nature* 454: 1065–1071.
WECD (The World Commission on Environment and Development, also known as the Brundtland Commission). 1987. *Our common future.* Oxford: Oxford University Press.

Chapter 2
The Growing Gap Between Emerging Technologies and the Law

Gary E. Marchant

It is change, continuing change, inevitable change, that is the dominant factor in society today. No sensible decision can be made any longer without taking into account not only the world as it is, but the world as it will be....
– Isaac Asimov

Change is inevitable, except from vending machines.
– Woody Allen

A series of concurrent technological revolutions are rapidly transforming economic, social and personal domains, now and even more so in the imminent future (Roco and Bainbridge 2003; Garreau 2005). These current and pending emerging technological revolutions include information technologies, communication technologies, nanotechnologies, biotechnology, regenerative and reproductive medicine, robotics, neuroscience, surveillance technologies, and synthetic biology. Perhaps even more important than the degree and breadth of these technological changes considered individually or collectively is the exponential pace at which the successive waves of technical change are washing over us (Kurzweil 2005).

In contrast to this accelerating pace of technology, the legal frameworks that society relies on to regulate and manage emerging technologies have not evolved as rapidly, fueling concerns about a growing gap between the rate of technological change and management of that change through legal mechanisms (Moses 2007). Increasingly, the traditional legal tools of notice-and-comment rulemaking, legislation and judicial review are being left behind by emerging technologies, struggling to cope with even yesterday's technologies.

The consequence of this growing gap between the pace of technology and law is increasingly outdated and ineffective legal structures, institutions and processes to regulate emerging technologies. The two basic options for addressing this problem are (i) to slow or stop the pace of scientific progress; or (ii) to improve the capacity

G.E. Marchant (✉)
Arizona State University, Tempe, AZ, USA
e-mail: gary.marchant@asu.edu

G.E. Marchant et al. (eds.), *The Growing Gap Between Emerging Technologies and Legal-Ethical Oversight*, The International Library of Ethics, Law and Technology 7,
DOI 10.1007/978-94-007-1356-7_2, © Springer Science+Business Media B.V. 2011

of the legal system to adapt to rapidly evolving technologies (even if this means departing from traditional forms of legal regulation into broader forms of governance, as discussed below). History indicates that the first option is highly unlikely, especially with technologies that have significant economic, psychological, or military value. Therefore, implementation of the second option becomes critical and is the focus of this introductory chapter. As Benjamin Cardozo noted almost half a century ago, "with new conditions there must be new rules" (Cardozo 1960, p. 137).

This chapter first summarizes evidence of the accelerating pace of science and technology, followed by evidence of the lagging efforts of law to keep up with emerging technologies. It then briefly describes some basic approaches and options to make the law more dynamic and responsive to accelerating technologies.

2.1 Accelerating Technology

Over the past half-century, we have witnessed the discovery of the structure of DNA, the mapping of the human genome, the development, refinement, and widespread adoption of computing systems, the advent of new communication and information technologies that are fundamentally changing economic and social relations, and the prevalent use of new technologies such as nanotechnology and neuroimaging that were not widely known just a decade ago.

Various statistical measures give a sense of the accelerating pace of science and technology. For example, one analysis concluded that over the last 250 years, the number of scientific journals has doubled approximately every 15 years and the number of "important discoveries" has doubled every 20 years (Price 1986; Tuomi 2003). Further, a study performed at the University of California-Berkeley showed that the total information in the world doubled from 1999 to 2002 (Lyman and Varian 2003). It has been estimated that we have created more scientific knowledge in the past four decades than was created in the previous 5,000 years (Garreau 2001). These exponential increases in information may be correlated with the available scientific workforce, which expanded dramatically and in some cases, exponentially. The number of engineers in the United States has doubled approximately every ten years (Price 1986; Tuomi 2003) and it has been estimated that 90 percent of all scientists who have ever lived are alive today (Garreau 2001).

The exponential growth of technology is demonstrated by several examples, one of which is Moore's Law, a 1965 observation of Gordon Moore, who subsequently co-founded Intel. Essentially, the observation suggests that at the current rate of technological development, the number of transistors that can fit on an integrated circuit (and hence, computing power) will double every 12 months (Moore 1965). (Moore revised this law in the 1970s to suggest that computing power would double every 18–24 months instead). Despite the exponential growth projected by Moore's Law, the rate of computer development has kept pace with these ambitious projections for nearly 50 years (Lundstrom 2003). It has been postulated that with the acceleration of Moore's law, transistors will reach their limit on "smallness" within the next decade; however, another 30 years of exponential progress is attainable and likely due to the rapidly increasing rate of advancement of technologies such as silicon nanoelectronics and quantum computing (Lundstrom 2003). Exponential improvements have also been demonstrated for computational power and for computer processing speed. Specifically, the value of $1,000 of computational power has doubled every two years through five paradigm shifts and most recently (the past 30 years) has doubled every year (Jurvetson 2004). Over the next 20 years, technological growth will be equal to that of the entire twentieth century (Jurvetson 2004). Further, desktop computer processing speed has seen exponential growth (Berndt et al. 2000).

Another area in which rapid technological advancement can be demonstrated is the biological sciences. An observation known as Monsanto's Law supports the theory of the increase of biological technology at an exponential rate. This law is derived from an observation made by the Monsanto Corporation in 1997: "The ability to identify and use genetic information is doubling every 12–24 months. This exponential growth in biological knowledge is transforming agriculture, nutrition and healthcare in the emerging life sciences industry" (Brand 1999).

The technology for DNA sequencing has also been improving at an exponential rate (Carlson 2003), with the cost of DNA sequencing dropping by a factor of 100,000-fold over the past decade (Carr 2010). Further, the growth of DNA sequence data that has been added to Genbank, a database of sequences, has been noted to be exponential as well. The amount of DNA sequence data has increased from 606 sequences in 1982 to 108,431,692 in 2009 (GenBank 2010).

Another area of biological technology that has advanced and evolved rapidly in the twentieth century is that of medical product innovation. In particular, the twentieth century has witnessed dramatic advances in pharmaceuticals, materials science, medical device engineering, and biologics. The advances in these individual technologies have culminated in the creation of an entirely new area of innovation, that of the technology of combination medical products (Bartlett Foote and Berlin 2005). Initially, combination medical products involved the addition of an existing pharmaceutical agent to a medical device. One example of an early combination medical product is the addition of a steroid drug to a pacemaker electrode in order to promote healing and decrease scarring of cardiac tissue. More recently, however, combination product innovation involves the addition (to medical devices) of agents that regenerate tissue (LaForte 2004; Bartlett Foote and Berlin 2005), target cells for delivery of gene sequences (Shea and Houchin 2004; Bartlett Foote and Berlin 2005), and detect illnesses in the body and target delivery of drugs and biologics to the illness locale (Miller 2003; Bartlett Foote and Berlin 2005). These innovative new combination products are the result of the rapidly increasing growth and advancement of the fields of tissue engineering, gene therapy, and nanotechnology, respectively (Bartlett Foote and Berlin 2005).

In the last 30 years, exponential advancement has also been demonstrated in noninvasive brain scanning technology. Specifically, exponential improvements in resolution of neuronal features have been demonstrated for computed tomography (CT) brain scanners (Trajtenberg 1990; Kurzweil 2003). In 1972, resolution capability was approximately 4.0 millimeters, and by 1999, resolution had decreased (improved) to approximately 0.06 mm (Kurzweil 2003).

Further, exponential growth has been demonstrated by advancements in internet connectivity (Jurvetson 2004) and in number of internet hosts (Internet Systems Consortium 2008). Specifically, the number of internet hosts has risen from 213 in 1981 to 541,677,360 in 2008 (Internet Systems Consortium 2008).

The field of nanotechnology has exhibited exponential growth as well. The number of nanotechnology science citations has increased from approximately 400 in 1990 to greater than 10,000 in 2002, with a doubling time of 2.4 years. As a result of increasing research and innovation in this particular area, the number of US nanotechnology-related patents has risen from approximately 45 in 1990 to approximately 500 in 2002 (http://www.etcgroup.org/documents/TheBigDown.pdf). Between 1997 and 2004, the number of patents granted in the area of nanotechnology has increased by a factor of three (Jurvetson 2004). As a result, IBM for example, has more lawyers involved in nanotechnology than engineers (Jurvetson 2004).

Finally, a review of patent applications indicates an increasing technological growth. Between 1995 and 2005 a total of over 1.6 million patent applications have been filed worldwide with an average annual growth rate of 4.7% (WIPO 2007). Nor has the growth rate been linear; between 1883 and 1959 the growth rate was a meager 1.99%, whereas between 2004 and 2005 it was 7% (WIPO 2007). A closer examination reveals that certain technical fields are growing at an even greater rate; between 2000 and 2004, medical technology (+32.2% increase in patent applications filed), audio-visual technology (+28.3%), and information technology (+27.7%) were the three fastest growing technical fields (WIPO 2007). Even when controlling for gross domestic product in the United States, the data show that there was a 6.3% increase in patent filings by residents in a one-year period from 2004 to 2005 (WIPO 2007).

2.1.1 Pace of Law vs. Pace of Science and Technology: Can Law Stay Current?

With the increasingly rapid progression of science and technology, a major challenge that arises is the capability of legislation, regulation, and judicial case law (collectively "law") to keep pace with rapidly developing science and technologies (Moses 2007). As far back as 1986, the US Office of Technology Assessment (OTA) noted that "[o]nce a relatively slow and ponderous process, technological change is now outpacing the legal structure that governs the system, and is creating pressures on Congress to adjust the law to accommodate these changes" (OTA 1986). A subsequent analysis from the RAND Corporation echoed this concern: "We see a growing divergence between time cycles of government and those of technology development. Quite simply, this presents government operations with a Hobson's choice: Either live within a shorter response time and run the concomitant risk of ill-considered actions (or inactions) or see government input become less relevant and assume reduced stature" (Popper 2003, p. 86).

The pacing problem facing the legal system has at least two dimensions. First, many existing legal frameworks are based on a static rather than dynamic view of society and technology. A textbook example of a legal provision that fails to anticipate likely change involves the ozone non-attainment provisions of the 1990 Amendments to the US Clean Air Act. After Congress had repeatedly failed in previous amendments to force states to comply with health-based ambient air quality standards, Congress determined in the 1990 amendments to hold states' feet to the fire by imposing a graduated series of progressively more onerous mandatory requirements scaled to the degree of non-compliance and the length of time the state needed to attain compliance. The problem was that Congress tied this graduated scale of requirements to the extent of non-attainment with the existing ozone standard (0.12 ppm averaged over one hour), with no flexibility or anticipation that the EPA ozone standard may change. In fact, the statute required EPA to review and revise as appropriate its ozone standard every five years, and so not unexpectedly, EPA significantly revised its ozone standard seven years later in 1997, changing both the stringency and form of the standard. The new attainment requirements no longer made sense in view of the revised standard, and extensive litigation and regulatory revisions resulted to try to resolve the mess made by Congress' failure to recognize that the world would not stand still.

Second, legal institutions are slowing down with respect to their capacity to adjust to changing technologies. This problem applies across the board to legislatures, regulatory agencies, and the courts. The *legislative* process is notoriously slow, with federal and state legislatures only capable of addressing a small subset of the plethora of potential issues before them in any legislative session. Issues are often not addressed on the basis of their importance, but rather as a function of headlines and perceived political urgency and expediency. Thus, a given issue may only be addressed by the legislature during an infrequent "window" when various factors combine to elevate the issue to the front of the priority line (Kingdon 1995). It often takes some type of crisis or disaster to open this legislative window and "shock" the legislature into taking legislative action (Kahn 2007). Once the legislative body has acted on an issue during the window of opportunity, it may be years or even decades before it revisits the issue, creating the risk of outdated legislation that remains in effect simply as a reflection of legislative inertia. For example, every US environmental statute is now past (in many cases well past) its stated reauthorization time (Campbell 2008). Statutes also fail to adapt because of political gridlock, where legislators agree that an existing statute is out-of-date, but cannot agree on how it should be changed, resulting in the prolonged life of an outdated statute. The legislative process is also heavily determined by the scale of relevant political and ideological structures, which in many cases are smaller than the increasingly global technological enterprise.

Regulatory processes by federal and state agencies have become slower at the same time that science and technology are speeding up. In the United States, regulatory agencies are required, both to meet legislatively imposed requirements and to survive judicial review, to undertake an ever-increasing burden of analytical requirements to support their regulatory decisions. The increasing complexity of the European Union regulatory processes similarly slows regulatory initiatives. As issues involving technology become more complex, more stakeholders become involved in regulatory processes, further slowing the potential for rapid regulatory action. These and other requirements have resulted in what is referred to as the "ossification" of rulemaking (McGarity 1992; Pierce 1995), whereby promulgation of new regulations becomes increasingly delayed and difficult, resulting in ineffective and out-dated regulations. Not only do agencies fail to adopt timely regulations to address new issues and problems (e.g., nanotechnology), but they fall into "rulemaking ruts" where they fail to update existing regulations in response to new scientific and technological knowledge (Blais and Wagner 2008). Although there remains empirical uncertainty and disputes about some aspects of the extent and causes of the ossification of rulemaking (e.g., Johnson 2008; Jordan 2000), there is no question that the evidentiary and legal burdens on agencies seeking to promulgate regulations has increased dramatically over time, contributing to a notable slowdown in rulemaking activities by some agencies (e.g., OSHA) and an apparent retreat from rulemaking altogether by other agencies (e.g., NHTSA) (Mashaw and Harfst 1991; Blais and Wagner 2008; McGarity et al. 2010).

The system of *judicial* case-law is deliberately structured to provide a conservative brake on rapid change in order to provide stability and predictability in the legal system. Thus, common law courts adhere to precedent, under the doctrine of stare decisis, often following the legal principles and holdings set down in cases decided decades or even centuries earlier (albeit with some flexibility to depart from such historical decisions in light of new facts, laws, and social views). The process of litigation is also often lengthy, as a single case can take many years to progress through the process from filing of a complaint to a final appellate decision, further increasing the likelihood that a judicial opinion might be outdated even at the time it is issued.[1]

These dynamics of legislative, regulatory and judicial legal actors all suggest that the law may have problems keeping pace with exponentially changing technologies. There are at least anecdotal examples of such a problem. In the Microsoft antitrust case in the United States, for example, the D.C. Circuit Court of Appeals noted that it was being asked to judge alleged anticompetitive conduct of Microsoft that took place six years earlier (US v. Microsoft 2001, at 49). The court expressed concern

[1] A (hopefully) trivial yet illustrative example of the slow pace of courts is a legal action in the European Court of Human Rights to stop the start up of the Large Hadron Collider in Europe because of an alleged risk it could start a runaway reaction that could destroy the earth. After the court denied an interim order to delay the experiment, a news report quoted a court spokesperson as saying it could "take several years" to decide the merits of the case, leading the reporter to caustically remark "[s]o, if a black hole is swallowing up the Earth by 2012, we might have the consolation of knowing it was illegal, at the conclusion of an apocalyptic version of Jarndyce v Jarndyce." (Warner 2008).

that "six years seems like an eternity in the computer industry," and "[b]y the time a court can assess liability, firms, products, and the marketplace are likely to have changed dramatically." (Id.). The court noted "the enormous practical difficulties" that resulted from the slow response of the legal system relative to the rapid pace of technology that, in this case, left the court evaluating what should be done about an earlier generation of now-obsolete software.

Another example of the legal system responding too slowly to changes in science and technology is the Delaney Clause, a 1958 amendment to the Food, Drugs, and Cosmetic Act sponsored by Congressman James Delaney of New York. The Delaney Clause prohibited any food additive that was "found to induce cancer in man, or, after tests, found to induce cancer in animals." At the time the clause was enacted, carcinogens were viewed as relatively rare substances that could be completely eliminated from the human diet. Soon thereafter, though, evolving scientific knowledge suggested that at least half of all chemicals could cause some form of cancer in animal tests at very high doses, and that almost every food additive and most "natural" foods contained some level of potential carcinogens at trace levels. Regulatory agencies such as FDA and EPA attempted to circumvent the harsh, extreme language of the Delaney Clause by suggesting that additives with trivial cancer risks should be exempted, but were repeatedly rejected by the courts that insisted only Congressional action could change the outdated assumptions underlying the Delaney Clause (Merrill 1988). It was not until 1996 that Congress finally stepped in to update the statute, decades after it was known to be scientifically obsolete and untenable in practice.

In some cases, lethargic development of new legislation or adaptation of existing legislation in response to scientific discovery or development can impede research and innovation, resulting in blocking of new technology. An example of this dilemma is the US patent system. While the purpose of patent law is to promote the progress of useful arts by providing inventors with incentives to innovate, thereby driving technological advancement, is has been suggested recently that US patent policy may, in fact, be impeding innovation (Mireles 2005; Jaffe and Lerner 2004). In particular, patent law has failed to recognize that different emerging technologies may require unique patenting rules, thereby continuing to apply an increasingly obsolete "one size fits all" policy to vastly different technologies (Thurow 1997; Burk and Lemley 2003).

Other examples of new technologies that have developed in a legal void with no comprehensive regulatory framework include embryonic stem cell research, artificial reproductive technologies (ART), preimplantation genetic screening, genetic testing, new surveillance technologies, and privacy on the internet. According to the co-chair of an American Bar Association committee studying the need for regulation of ARTs, "[w]ith each advancement in technology, the law grows further behind" (Baker 1999).

In addition to these anecdotal examples, there appears to be a common sentiment among experts in a variety of different technologies that regulatory systems are not keeping pacing with the rapidly developing technology. For example, such expert statements in the field of nanotechnology include:

- Innovation in the field of nanotechnology development is far ahead of the policy and regulatory environment, which is fragmented and incomplete at both the national and international levels. (Roco and Renn 2007)
- We have moved into . . . a . . .world dominated by rapid improvements in products, processes, and organizations, all moving at rates that exceed the ability of our traditional governing institutions to adapt or shape outcomes. If you think that any existing regulatory framework can keep pace with this rate of change, think again. (Rejeski 2004)
- [L]aw, regulations, and policy are going to have to take a giant leap if they are to keep up with the pace of nanotechnology development. (Kelly 2008)
- Currently, governments are not able to set up or modify comprehensive regulatory structures quickly enough to match the pace of innovation and product introduction. (IRGC 2007)
- The slow movement to regulate nano in the face of legitimate yet manageable risks is an example of the broader social issue of how regulation is sorely lagging behind advances in technology generally. (Laws 2008)

Likewise in the life sciences, commentators express concern about the ability of law to keep pace with new scientific developments:

- [B]iomedical technologies are quickly outpacing the development of appropriate policies to inform the decision-making of researchers and the general public on many issues, including genetic testing, medical privacy, genetic discrimination and others. (Terry 2001)
- Science and technologies are outpacing the development of appropriate policies for decisionmaking. Genetic testing, medical privacy, genetic discrimination and others are some of the issues we face without having the right policies in place. (Eibert 2002)
- Although advances in genomics hold the potential for a range of preventive medical interventions of great value, risks to patients are also emerging. Laboratory regulation and accreditation measures have not kept pace with the growing demand for genetic tests. (WHO 2008)
- [R]egulatory institutions have not kept pace with our rapid technological advance. Indeed, there is today no public authority responsible for monitoring or overseeing how these [reproductive] technologies make their way from the experimental to the clinical stage, from novel approach to widespread practice. There is no authority, public or private, that monitors how or to what extent these new technologies are being or will be used, or that is responsible for attending to the ways they affect the health and well-being of the participants or the character of human reproduction more generally. (President's Council on Bioethics 2004)
- The synthetic life sciences seem to have emerged from nowhere, and their potential uses and misuses have taken the scientific and regulatory community by surprise . . .. [I]t is a reminder of how scientific development might leave moral, social, legal discourse in its wake. . .. (Samuel et al. 2009)

The same observation of lagging legal oversight is frequently made for information and communication technologies:

- As applied to the Internet, the traditional legal issues of property rights, commerce and trade, national sovereignty, and international remedy, are being examined in this new context. The old rules do not well apply, for this technology has outpaced the law. (Newman 2003)
- [E]lectronic surveillance law ... has failed to keep pace in adapting to new technologies, and ... provides for insufficient judicial and legislative oversight. (Solove 2004)
- Legislation ... is a slow process, often at pains to keep pace with rapid technological advance .... Technologies, after all, change faster than laws can. (Garfinkel 2002)
- The Internet has given the government powerful 21st-century tools for invading people's privacy and monitoring their activities, but the main federal law governing online privacy is a 20th-century relic. Adopted in 1996, it has had trouble keeping up with technological advances and is now badly out of date. (NY Times 2010)

This concern that law is badly trailing rapidly evolving technologies has been repeated literally dozens of time by experts working on dozens of technologies (See Moses 2007 for examples). Of course, there may be examples where it is preferable to delay writing new laws for regulating emerging technologies until the direction and risks of the new technology have been better determined. Premature regulation of a developing technology could result in poorly aligned regulations that inappropriately "lock in" inferior technological choices (Fong 2001). For example, attempts to regulate nanotechnology at the present time would likely be premature because the future direction and risks of this emerging technology (and, indeed, even an appropriate definition given the breadth of the term) are so uncertain.

Moreover, one of the key functions of law is to serve as a stabilizing mechanism to restrain rapid change based on human dynamics and preference shifts (Van Alstine 2002). Former Supreme Court Chief Justice Warren Berger once wrote: "It should be understood that it is not the role and function of the law to keep fully in pace with science" (Berger 1967). In some cases, then, the slow pace of legal evolution may provide important benefits by preserving stability and certainty. Finally, there may be cases where the law is too far in front of science rather than trailing behind it. Such an example might be the early litigation over silicone breast implants, where judges and juries were required to decide the safety of the implants before adequate scientific data on the safety of the devices had been generated (Jasanoff 1995).

These counterexamples, however, do not undermine the suggestion that, especially with foundational technology systems such as biotechnology, nanotechnology, and information and communication technologies, there is a substantial generic and novel problem caused by the inability of law to keep pace with rapidly developing science and technologies. As Waldmeir (2001) noted, "Throughout history, technology has always outpaced the law. In the end, the law catches up. But now, faced with technologies of unprecedented power, there is a risk . . .. that things could be different." Lyria Bennett Moses (2007) has identified four potential problems that may result from the failure of law to keep pace with technology, including: (i) the failure to impose appropriate legal restrictions and precautions to control the risks of new technologies; (ii) uncertainties in the application of existing legal frameworks to new technologies; (iii) the potential for existing rules to either under– or over-regulate new technologies; and (iv) the potential for technology to make existing rules obsolete.

Legal and regulatory systems have generally been oblivious to the growing lag between legal oversight mechanisms and the rapid pace of emerging technologies. One notable exception is the European Union's "Better Regulation" initiative, an integrated series of undertakings to improve the European regulatory system (European Commission 2010). A key objective of this initiative is "to identify overlaps, gaps, inconsistencies, obsolete measures, and potential for reducing regulatory burdens" without sacrificing the level of protection provided by the regulatory system (Commission of the European Community 2009, at 3). Even this initiative, however, has not achieved major successes in addressing the pacing problem, other than reducing some regulatory redundancies and overlaps.

A variety of potential legal mechanisms might provide a more flexible and adaptive regulatory system that can avoid or minimize some of the problems identified above. Some possibilities include:

1. *Innovations for Expediting Rulemaking*: As regulatory rulemaking has become increasingly lengthy and burdened with analytical and procedural requirements, regulatory agencies have recently begun exploring possible approaches for streamlining rulemaking. An example of such a measure is direct final rulemaking, whereby an agency issues a final rule without going through public comment, but will withdraw the rule and go through full notice-and-comment rulemaking if substantial public comments in opposition to the direct final rule are received (Levin 1995). This and other recent innovations attempted by federal or state regulatory agencies or proposed by administrative law scholars will be evaluated as strategies for keeping regulation more up-to-speed with science and technology.
2. *Self-Regulation or "Cooperative" Regulation*: In recent years, regulatory agencies such as the US Environmental Protection Agency (EPA) and the Occupational Safety and Health Administration (OSHA) have increasingly relied on "cooperative regulation" approaches under which industry essentially self-regulates under the supervision of the relevant federal agency (Pedersen 2001). The European Union and its member states are likewise increasingly recognizing and relying on self-regulatory approaches to fill the gaps and supplement traditional regulatory frameworks (Falkner et al. 2005). A prominent example is the covenant system of environmental regulation in the Netherlands, which is based on voluntary agreements between industry, government, and environmental organizations (Allenby 1999). Such self-regulatory approaches have the

benefit of achieving more expedited results without having to comply with all the formalities of official regulation, but also raise concerns about accountability and the lack of opportunity for public participation in regulatory decisions (Caldart and Ashford 1999).

3. *Issue-Specific Legislative Initiatives*: Congress recently enacted new statutes in several subject matters with the express purpose of making the statutes more current with changing technologies. Examples include the FDA Modernization Act of 1997 and the Digital Millennium Copyright Act of 1998. The process, effectiveness, and feasibility of such issue-specific statutory approaches remain to be evaluated.
4. *Specialized Courts*: In recent years, states such as Michigan and Maryland have taken steps to establish technology or cyber-courts to provide speedier and more sophisticated legal decisions in cases involving rapidly changing technologies (Maryland Business and Technology Court Task Force 2000; Ponte 2002). The feasibility and design of these specialized technology courts need to examined with respect to their potential for providing speedier legal resolution of technological disputes.
5. *Sunset Clauses*: One mechanism that legislatures can use to prevent statutes from becoming out-dated is to insert a sunset clause that results in the automatic expiration of the legislation after a given time-period (Mooney 2004). As the President's Council on Bioethics stated when recommending a four-year moratorium on therapeutic cloning rather than a permanent ban, a temporary ban that expires after a given period "make[s] it impossible for either side to cling to the status-quo," because the sunset of the legislation would force legislators to revisit the issue de novo (President's Council on Bioethics 2002).
6. *Periodic Reviews*: Some regulatory programs build in mandatory periodic reviews to assess the current status of the problem being regulated and to evaluate whether programmatic adjustments are warranted. An example of such a requirement is California's adoption of the zero emission vehicle mandate in 1990, scheduled originally to commence implementation in model year 1998. Recognizing the technology forcing challenge this mandate imposed on manufacturers, California regulators provided for a biennial program review that resulted in numerous delays and revisions to the applicable requirements. This periodic review provision resulted in considerable program instability, uncertainty and strategic behavior, but also allowed the program to adjust to changing expectations about the feasibility of zero emission vehicles.
7. *Independent Institutions*: Another strategy might be to create and delegate decision-making authority over an issue to a new free-standing commission or other institution that can make efficient, speedy adjustments in policy as a technology evolves. Such an entity would be less burdened by the political and bureaucratic restrictions that apply to existing political institutions. An example of such an institution might be the Commission on Base Closure and Realignment established by Congress in 1991 to determine which military bases should be closed which, according to at least one analysis, was able

to successfully protect professional norms from political pressures and make efficient decisions on the merits of the issue (Koven 1992). Independent institutions overseeing emerging technologies at the international level may also provide appropriate flexibility and adaptability. A good example is the Internet Corporation for Assigned Names and Numbers (ICANN), an independent entity given responsibility for managing the internet domain system and the many related technical and policy issues. According to ICANN, it's independent status gives it the flexibility to respond quickly to fast moving technological change: "ICANN's independence enables rapid response to changes within the commercial, technical and geopolitical landscape of the Internet and DNS. While rapid and flexible, the ICANN process also requires and considers input from all interested and affected constituencies." (ICANN undated).

8. *Adaptive Management*: Adaptive management was developed initially in ecology to provide an iterative approach for regulating complex ecosystems subject to frequent change. The premise of such adaptive management approaches was to use continual feedback to adjust policy in parallel with changing facts. In recent years, legal scholars have proposed that law could adopt a similar adaptive management approach (Ruhl 1997).
9. *Principles Based Regulation*: "Principles Based Regulation" is a new approach to regulation that involves the promulgation of general principles of expected behavior rather than detailed prescriptive rules. Regulated parties are then expected to implement the general principles in their own regulatory programs. The reliance on more general principles rather than detailed rules can provide greater flexibility for adjustments in response to new developments without the need to promulgate new regulations. This new approach has primarily been attempted in the financial services sector particularly in the United Kingdom to date, but a broader application has been proposed to provide a more adaptive regulatory system.

References

Allenby, Braden R. 2005. *Reconstructing earth: Technology and environment in the age of humans.* Washington, DC: Island Press.

Allenby, Braden R. 1999. *Industrial ecology: Policy framework and implementation.* Upper Saddle River, NJ: Prentice-Hall.

Baker, Debra 1999. Catching up to science. *American Bar Association Journal* 88 (Dec.): 88.

Bartlett Foote, Susan, and Robert J. Berlin. 2005. Can regulation be as innovative as science and technology? The FDA's regulation of combination products. *Minnesota Journal of Law, Science & Technology* 6: 619, 620–623.

Berger, Warren E. 1967. Reflections on law and experimental medicine. *UCLA Law Review* 15: 436–442.

Berndt, Ernst R., E.R. Dulberger, and N.J. Rappaport. 2000. Price and quality of desktop and mobile personal computers: A quarter century of history, 17 July 2000).

Blais, Lynn E., and Wendy E. Wagner. 2008. Emerging science, adaptive regulation, and the problem of rulemaking ruts. *Texas Law Review* 86: 1701–1739.

Brand, Stewart. 1999. *The Clock of the long now.* New York, NY: Basic Books.

Burk, Dan, and Mark Lemley. 2003. Policy levers in patent law. *Virginia Law Review* 89: 1575, 1575–1696.

Caldart, Charles C., and Nicholas A. Ashford. 1999. Negotiations as a means of developing and implementing environmental and occupational health and safety policy. *Harvard Environmental Law Review* 23: 141–202.

Campbell, Bradley M. 2008. Landmarks and land mines. *The Environmental Forum* (Nov./Dec.): 30–35.

Cardozo, Benjamin N. (1960). *The nature of the legal process* (Yale Paperbound ed 1960).

Carlson, Robert. 2003. The pace and proliferation of biological technologies. *Biosecurity and Bioterrorism: Biodefense Strategy, Practice, and Science* 1: 203, 203–14.

Carr, Geoffrey 2010. Biology 2.0. *The Economist* (June 19, Supp: 1–3).

Commission of the European Communities. (2009). Communication from the Commission to the European Parliament, the Council, the European Economic and Social Committee and the Committee of the Regions, Third Strategic Review of Better Regulation in the European Union, COM(2009) 15 final.

Eibert, Mark D. 2002. Human cloning: Myths, medical benefits and constitutional rights, *Hastings Law Journal* 53: 1097.

European Commission. 2010. Better Regulation. available at http://ec.europa.eu/governance/better_regulation/key_docs_en.htm#_br.

Falkner, Gerda, Oliver Treib, Miriam Hartlapp, and Simone Leiber. (2005). *Complying with Europe: EU harmonisation and soft law in the member states*. Cambridge: Cambridge University Press.

Kong Fong, I. 2001. Law and new technology: The virtues of muddling through. *Yale Law & Policy Review* 19: 443, 443–461.

Garfinkel, Simson. Oct 2002. An RFID bill of rights. *Technology Review* 35 *available at* http://www.simson.net/clips/2002/2002.TR.10.RFID_Bill_Of_Rights.pdf.

Garreau, Joel. 2001. *Science's mything links: As the boundaries of reality expand, our thinking seems to be going over the edge*. Washington Post Cl (July 23).

Garreau, Joel. 2005. *Radical evolution*. New York, NY: Random House).

Genbank 2010. GenBank Overview: What is Genbank? available at http://www.ncbi.nlm.nih.gov/genbank/. Accessed 14 July 2010.

ICANN (undated). ICANN Factsheet. available at http://www.icann.org/en/factsheets/fact-sheet.html.

International Risk Governance Council (IRGC). 2007. *Nanotechnology risk governance*. IRGC: Geneva.

Internet Systems Consortium, *ISC Domain Survey: Number of Internet Hosts*. 2008. http://www.isc.org/index.pl?/ops/ds/host-count-history.php.

Jaffe, Adam B., and Josh Lerner.2004. *Innovation and Its discontents: How our broken patent system is endangering innovation and progress, and what to do about it*. Princeton, NJ: Princeton University Press.

Jasanoff, Sheila. 1995. *Science at the bar: Law, science, and technology in America*. Cambridge, MA: Harvard University Press.

Johnson, Stephen M. 2008. Ossification's demise? An empirical analysis of EPA rulemaking from 2001–2005. *Environmental Law* 38: 767–792.

Jordan, W.S., III. 2000. Ossification revisited: Does arbitrary and capricious review significantly interfere with agency ability to achieve regulatory goals through informal rulemaking? *Northwestern University Law Review* 94: 393–450.

Jurvetson, Steve. 2004. *Transcending Moore's law with molecular electronics and nanotechnology. Nanotechnology Law & Business* 1: 70–90.

Kahn, Matthew E. 2007. Environmental disasters as risk regulation catalysts? The role of Bhopal, Chernobyl, Exxon Valdez, Love Canal, and Three Mile Island in Shaping US environmental law. *Journal of Risk and Uncertainty* 35: 17–43.

Kelly, James J. 2008. President of the Food and Drug Law Institute, *quoted in* David J. Hanson, *FDA Confronts Nanotechnology*, Chemical & Engineering News, 17 March 2008, at 32–34.

Kingdon, John W. 1995. *Agendas, alternatives, and public policies,* 2nd ed. New York, NY: Addison Wesley Educational Publishers.

Koven, Steven G. 1992. Base closings and the politics-administration dichotomy revisited. *Public Administration Review* 52: 526–531.

Kurzweil, Ray. 2003. Exponential growth an illusion?: Response to Ilkka Tuomi, essay (September 23) http://www.kurzweilai.net/meme/frame.html?m=1.

Kurzweil, Ray. 2005. *The singularity is near: When humans transcend biology.* London: Penguin.

LaForte, Amy J. 2004. Bone Morphogenetic Protein Combination Products and Orthopedic Repair, in Nat'l Research Council. *Proceedings From The Workshop On Science-Based Assessment: Accelerating Product Development Of Combination Medical Devices* 15 (Bonnie A. Scarborough ed).

Laws, Elliott P. 2008. Regulators facing a brave new world, The environmental forum. July/August 2008, at 14.

Levin, Ronald M. 1995. Direct final rulemaking, *George Washington Law Review* 64: 1, 1–34.

Lundstrom, Mark. 2003. Moore's Law forever? *Science* 299: 210, 210–211.

Lyman, Peter, and Hal R. Varian. 2003. *How much information.* available at http://www2.sims.berkeley.edu/research/projects/how-much-info-2003/ on 2 Nov. 2005.

Maryland Business and Technology Court Task Force, Final Report. 2000. available at http://www.courts.state.md.us/finalb&treport.pdf.

Mashaw, Jerry L., and David L. Harfst. 1991. *The struggle for auto safety.* Cambridge, MA: Harvard University Press.

McGarity, Thomas O. 1992. Some thoughts on deossifying the rulemaking process. *Duke Law Journal* 41: 1385, 1385–1462.

McGarity, Thomas, Rena Steinzor, Sidney Shapiro, and Mathew Shudtz. 2010. Workers at risk: regulatory dysfunction at OSHA, Center for Progressive Reform White Paper #1003.

Merrill, Richard A. 1988. FDA's implementation of the delaney clause: Repudiation of congressional choice or reasoned adaptation to scientific progress? *Yale Journal on Regulation* 5: 1, 1–88.

Miller, J. 2003. Beyond biotechnology: FDA regulation of nanomedicine. *The Columbia Science and Technology Law Review* 4: 1, 1–2.

Mireles, Michael S. 2005. The United States patent reform quagmire: A balanced proposal. *Minnesota Journal of Law, Science & Technology* 6: 709.

Mooney, Chris 2004. *A short history of sunsets. Legal Affairs* 67, 67–71. (Jan.–Feb. 2004), available at http://www.legalaffairs.org/issues/January-February-2004/story_mooney_janfeb04.msp.

Moore, Gordon E. 1965. Cramming more components onto integrated circuits. *Electronics* 38 (8): 114–117.

Moses, Lyria Bennett. 2007. Recurring dilemmas: The law's race to keep up with technological change. *University of Illinois Journal of Law, Technology & Policy* 2007: 239–285.

Newman, P. 2003. Trilateral seminar on science, society and the internet, opening statement by Judge Pauline Newman. A Dec. 14–16, conference sponsored by, inter alia, the NSF, available at www.law.gmu.edu/nctl/stpp/us_japan_pubs/internet-IIAIIB.pdf

New York Times. 2010. Dial-Up law in a broadband world (editorial). *NY Times*, April 9, at A18.

Office of Technology Assessment. 1986. *Intellectual property rights in an age of electronics and information.* Washington, DC: GPO.

Pedersen, William F. 2001. Contracting with the regulated for better regulations. *The Administrative Law Review.* 53: 1067, 1067–1138.

Pierce, Richard J. Jr., 1995. Seven ways to deossify agency rulemaking. *Administrative Law Review* 47: 59, 60.

Ponte, Lucille M. 2002. The Michigan Cyber Court: A bold experiment in the development of the first public virtual courthouse. *North Carolina Journal of Law & Technology* 4: 51, 51–91 (2002).

Popper, Steven W. 2003. Technological change and the challenges for 21st century governance. In *AAAS science and technology policy yearbook 2003*, eds. A.H. Teich et al., 83–103. Washington: American Association for the Advancement of Science.
President's Council on Bioethics. 2002. *Human cloning and human dignity: An ethical inquiry.* Washington, DC: GPO.
President's Council on Bioethics. 2004. Reproduction and responsibility: The regulation of new biotechnologies. Washington, DC: GPO.
de Solla Price, D.J. 1986. *Little science, big science ...and beyond.* New York, NY: Columbia University Press.
Rejeski, David. 2004. *The next small thing*. The environmental forum, March/April 2004, at 45.
Roco, Mihail C., and William Sims Bainbridge. 2003. *Converging technologies for improving human performance: Nanotechnology, biotechnology, information technology and cognitive science*. Dordrecht: Klewer Academic Publishers.
Roco, Mihail., and Ortwin Renn. 2007. International Risk Governance Council, Policy Brief: Nanotechnology Risk Governance – Recommendations for a Global, Coordinated Approach to the Governance of Potential Risks.
Ruhl, J.B. 1997. Thinking of environmental law as a complex adaptive system: How to clean up the environment by making a mess of environmental law. *Houston Law Review* 34: 933–1002.
Samuel, Gabrielle N., Michael J. Selgelid, and Ian Kerridge. 2009. Managing the unimaginable. *EMBO Reports* 10: 7–11.
Shea, Lonnie D., and Tiffany L. Houchin. 2004. Modular design of non-viral vectors with bioactive components. *Trends In Biotechnology* 22: 429.
Solove, Daniel J. 2004. Reconstructing electronic surveillance law. *George Washington Law Review* 72: 1264.
Terry, Sharon T. 2001. Prepared testimony before the Subcommittee on Oversight and Investigations of the Committee on Energy and Commerce, House of Representatives, 107th Congress, *Issues Raised by Human Cloning Research,* 28 Mar. 2001.
Thurow, Lester C. 1997. Needed: A new system of intellectual property rights. *Harvard Business Review* 75 (5): 94–103.
Trajtenberg, Manuel. 1990. *Economic analysis of product innovation: the case of CT scanners*, Cambridge, MA: Harvard University Press.
Tuomi, Ilkka. 2003. *Kurzweil, Moore, and Accelerating Change*, Joint Research Centre, Institute for Prospective Technological Studies, Working paper 27 (August 2003), http://www.meaningprocessing.com/personalPages/tuomi/articles/Kurzweil.pdf.
United States v. Microsoft Corp. 2001. 253 F.3d 34. (D.C. Cir.).
Van Alstine, Michael P. 2002. The costs of legal change. UCLA Law Review. 49: 789–870.
Waldmeir, Patti. 2001. Lawmakers Struggle to Keep Up, Financial Times, 2 Oct. 2001 at 13.
Warner, Gerald. 2008. If You're Reading This, Perhaps All is Well. *Daily Telegraph* (Sept. 10) at 21.
World Health Organization (WHO). 2008. *Quality & Safety in Genetic Testing: An Emerging Concern.* available at: http://www.who.int/genomics/policy/quality_safety/en/index.html.
WIPO (World Intellectual Property Organization). 2007. Patent Report: Statistics on Worldwide Patent Activity (2007 Edition). http://www.wipo.int/ipstats/en/statistics/patents/patent_report_2007.html#P173_14118.

Chapter 3
Ethical Challenges of Emerging Technologies

Joseph R. Herkert

I don't see anything wrong with human life being devalued if we have something better.
– Marvin Minsky (Anon. 2000)

Privacy is dead – get over it.
– Scott McNealy (Timmins 2006)

Ethicist Kristin Shrader-Frechette has noted (1997):

> Throughout history, technology...has opened new possibilities for actions. As a result, it has also raised new ethical questions....Most of these questions have not generated new ethical concepts; instead they have expanded the scope of existing ones. (p. 25)

The purpose of this chapter is to examine whether Shrader-Frechette's claim holds true for currently emerging technologies, including nanotechnology, neurotechnology, biotechnology, robotics, and advanced information and communication technology, or whether new ethical concepts are, indeed, called for. Before discussing ethical challenges posed by these technologies, however, it will be useful to focus on developments in two of these fields, namely humanoid robotics and pervasive computing.

3.1 Humanoid Robotics

Robots are becoming more and more prominent in our technological society including robot vehicles, industrial robots, space robots, personal robots, service robots, and robots for biomedical applications. A class of robots of particular interest is humanoid robots. Robots that look, think, and act like humans have long been the stuff of science fiction, but research on humanoids is beginning to bear fruit.

J.R. Herkert (✉)
Arizona State University, Tempe, AZ, USA
e-mail: joseph.herkert@asu.edu

G.E. Marchant et al. (eds.), *The Growing Gap Between Emerging Technologies and Legal-Ethical Oversight*, The International Library of Ethics, Law and Technology 7,
DOI 10.1007/978-94-007-1356-7_3,

According to the Idaho National Laboratory (http://www.inl.gov/adaptiverobotics/humanoidrobotics/),

> [H]umanoid Robotics includes a rich diversity of projects where perception, processing and action are embodied in a recognizably anthropomorphic form in order to emulate some subset of the physical, cognitive and social dimensions of the human body and experience. . . .Humanoids will interact socially with people in typical, everyday environments. We already have robots to do tedious, repetitive labor for specialized environments and tasks. Instead, humanoids will be designed to act safely alongside humans, extending our capabilities in a wide variety of tasks and environments.

Robotic applications (not all of which are humanoid) that are either commercially available or the subject of significant ongoing research include pets, toys, vacuum cleaners, lawn mowers, armed security guards, armed border sentries, construction and factory workers, maids, home monitors ("tele-presence robots"), companions for children, caretakers for the elderly and infirm, museum tour guides, and combat troops. There are currently more than 4,000 military robots deployed on the ground in Iraq (as well as unmanned aircraft). One observer has noted: "A robot does what it's told, and you'll be able to get them to advance in ways it's hard to get human soldiers to do. They don't have fear, and they kill without compunction." (Pasulka 2008) MIT researchers are working on robots that display human emotions. Of the nearly one million robots already in operation about half are in Asia, a third in Europe, and only one-sixth in North America. One Japanese researcher has even created a lifelike robot twin of himself.

Along with developments in robot technology, scientists, technologists, and ethicists are beginning to develop an ethics of robots. South Korea and Japan have already begun work on codes of ethics for the development of robots, both for the protection of humans *and* the protection of robots. Professional groups have begun to form around ethical issues, such as the Technical Committee on Roboethics of the Institute of Electrical and Electronics Engineers (IEEE), and other groups such as the European Robotics Network (Euron) have begun engaging in ethics and legislative activities. Humanoid robots pose a number of ethical dilemmas relating to such concepts as moral agency, free will, human identity, social roles, and potential marginalization of humans. Issues include consumer safety, product liability, and whether robots should/will ultimately have rights (as in the current case of debates over animal rights). Robots used in military roles will also raise a number of ethical questions including accountability for the robots' behavior. Perhaps trumping all of these concerns is the question of whether robots will become autonomous beings capable of making ethical decisions.

3.2 Pervasive Computing

The Centre for Pervasive Computing (Denmark) (http://www.pervasive.dk/) states that:

> Pervasive computing is the next generation computing environments with information & communication technology everywhere, for everyone, at all times. Information and communication technology will be an integrated part of our environments: from toys, milk cartons

> and desktops to cars, factories and whole city areas – with integrated processors, sensors, and actuators connected via high-speed networks and combined with new visualisation devices ranging from projections directly into the eye to large panorama displays.

Many such technologies were highlighted in the 2002 Steven Spielberg sci-fi film *Minority Report* starring Tom Cruise, but pervasive computing (also known as ubiquitous computing, ambient intelligence, and "everyware") is becoming reality. Pervasive computing is being enabled by the rapid convergence of advances in such fields as microelectronics, materials science, solid-state physics, nanotechnology, radiofrequency identification (RFID), wireless communication, and global positioning systems (GPS). Pervasive computing has the potential to impact every facet of our lives, including our homes, workplaces, schools, businesses, and entertainment venues. The social and ethical implications of these technologies include potentially significant environmental impacts (resource use, waste streams), health and safety impacts (non-ionizing radiation, psychological stress), and social impacts (digital divide, freedom of choice, information overload, privacy).

The early outlines of a pervasive computing environment are already beginning to emerge. There are more than 225 million cell-phone subscribers in the US, about 70% market penetration. More US adults now live in cell-phone only homes than in land-line only homes. In parts of Asia and Europe the penetration rates are even higher. The Apple iPhone 3G and its competitors incorporate the latest generation of wireless communication technology and GPS technology. Credit-card-size smart cards are used for controlling access to commercial applications, transportation systems, and buildings. Wearable computing technologies are used in military, medical, and commercial applications (e.g., wrist watch computers). RFID tags and/or implants are widely used to identify products, animals and people (including in US passports). Human RFID implants are commercially available and promoted for medical and security purposes.

3.3 Are Emerging Technologies Unique?

Humanoid robotics and pervasive computing are but two of many hundreds if not thousands of examples of emerging technologies. There is a tendency to shrug such developments off as business as usual, the inevitable result of technological progress. But emerging technologies seem to have characteristics that distinguish them from other technologies in significant ways.

Emerging technologies are sometimes referred to as "converging technologies" due to the fact, discussed in more detail later, that they overlap in many non-trivial ways. Humanoid robotics and pervasive computing, for example, both rely heavily on many of the same earlier developments in information and communication technology (e.g., sensors and wireless networking) and the utility of robots will be greatly increased by a pervasive computing environment that interacts with the robots.

A recent European Commission study on converging technologies (Nordmann 2004) identified a number of other general characteristics of emerging technologies that set them apart from earlier technological innovations and, in turn, make them ethically challenging:

- Embeddedness: the breaking down of "traditional boundaries between the self, nature and social environment, where the social environment includes people, groups of people, informal and formal institutions."
- Unlimited Reach: the promise of a "technological fix" for every problem.
- Engineering the Mind and the Body: from autonomous machines (e.g., killer robots) to transcending human limitations through human-machine interfaces.
- Specificity: diffusion of specifically targeted technology across an artificial environment (e.g., pervasive computing).

3.4 Who Should Do the Ethics?

If emerging technologies pose unique ethical challenges, who is in the best position to meet those challenges? As noted earlier in the discussion of robotics, the promoters of emerging technologies have begun to stake out the "moral ground" surrounding these novel technologies. For example, The Institute for Ethics and Emerging Technologies (IEET) (http://ieet.org/) (associated with the World Transhumanist Association) has as its mission:

> ...to become a center for voices arguing for a responsible, constructive approach to emerging human enhancement technologies. We believe that technological progress can be a catalyst for positive human development so long as we ensure that technologies are safe and equitably distributed. We call this a "technoprogressive" orientation.
>
> We aim to showcase technoprogressive ideas about how technological progress can increase freedom, happiness, and human flourishing in democratic societies. Focusing on emerging technologies that have the potential to positively transform social conditions and the quality of human lives – especially "human enhancement technologies" – the IEET seeks to encourage public policies for their safe and equitable use, and to cultivate academic, professional, and popular appreciation about their impacts.

Indeed, many of the promoters of emerging technologies see ethical action taking on new meanings, for example in military robots that can be programmed to follow the conventions of war (Arkin 2008) or in intelligent machines that transcend not only human intelligence as we know it but also human moral character (Hall 2007).

Of course, "technoprogressive" from one perspective may sound an awful lot like "technocratic" from another perspective. Rather than ceding this moral ground to those who promote converging technologies, it is prudent to consider what more traditional approaches to ethics have to offer.

3.5 Microethics and Macroethics in Engineering

During the past two to three decades, as engineering ethics has emerged as an academic subfield, several authors, including the ethicist John Ladd (1985), have noted that the content of engineering ethics encompasses multiple domains. The field can be viewed from three frames of reference – individual, professional, and social – that can be divided into "microethics," concerned with ethical decision making by individual engineers and the engineering profession's internal relationships, and "macroethics," referring to the profession's collective social responsibility and to societal decisions about technology (Herkert 2001). Microethical issues in engineering include such matters as designing safe products and not accepting bribes or participating in kickback schemes. Macroethics in engineering includes the social responsibilities of engineers and the engineering profession concerning societal issues such as sustainable development and the ethics of emerging technologies. This distinction can also be applied to other fields of applied ethics, such as research ethics (see table below).

Some microethical and macroethical issues in science and engineering

	Engineering practice	Scientific research
Microethics	Health and safety	Integrity of data
	Bribes and gifts	Fair credit
Macroethics	Sustainable development	Human cloning
	Emerging technologies	Stem cell research

Engineering ethics research and teaching to date have for the most part focused on microanalysis of individual ethical dilemmas in such areas as health and safety issues in engineering design, conflict of interest, representation of test data, whistle blowing, accountability to clients and customers, quality control, trade secrets, and gift giving (Herkert 2000) with little attention being paid to macroethics in engineering and still less to attempts at integrating microethical and macroethical approaches to engineering ethics. The melding of ethics and professionalism has significantly contributed to the development of engineering ethics concepts. At the same time, however, by overemphasizing issues internal to the profession, engineering ethicists have historically given short shrift to macroethical issues (O'Connell and Herkert 2004).

Recently engineering ethicists and engineering leaders alike have begun to turn their attention to macroethical issues by appealing directly to policy makers and to the engineering profession. For example, the National Academy of Engineering (NAE) recently conducted a workshop on emerging technologies and ethical issues in engineering that featured papers on environmental issues, nanotechnology, neurotechnology, and energy production and utilization, as well as sessions on engineering ethics research and education. The workshop was the initiative of then NAE President Bill Wulf who for several years has been urging the engineering profession to take seriously its responsibilities in the realm of macroethics:

> Several things have changed, and are changing, in engineering that raise macroethical questions. I'm going to talk only about the one that is closest to my professional experience—complexity. The level of complexity of the systems we are engineering today, specifically systems involving information technology, biotechnology, and increasingly nanotechnology, is simply astonishing. When systems reach a sufficiently high level of complexity, it becomes impossible to predict their behavior. It's not just hard to predict their behavior, it's *impossible* to predict their behavior. The question can't be answered by

> taking more things into account or thinking harder about the problem or using a new set of tools. At a certain threshold of complexity, it becomes impossible to predict all system behaviors. (Wulf 2003, p. 4)

The NAE went on to establish the Center for Engineering, Ethics and Society (CEES). In addition to assuming responsibility for the former Online Ethics Center in Engineering and Science (http://www.onlineethics.com/), the CEES's first major project was a workshop on Engineering Ethics, Social Justice and Sustainable Community Development held in October 2008. The workshop focused on social and engineering perspectives on engineering for social justice and sustainable development at the community level, and on identifying appropriate responses from engineering ethics research and education and from professional engineering societies.

3.6 Ethicists and Emerging Technologies

In parallel to the movement toward macroethical analysis in engineering ethics, a few ethicists in other relevant fields have begun to question the appropriateness of traditional approaches to ethics in light of emerging technologies. Noted computer ethicist James Moor, for example argues (2005) for the need for "better ethics for emerging technologies:"

> ...we are living in a period of technology that promises dramatic changes and in which it is not satisfactory to do ethics as usual. Major technological upheavals are coming. Better ethical thinking in terms of being better informed and better ethical action in terms of being more proactive are required. (p. 111)

Moor suggests that certain technologies, such as information technology, are "revolutionary" in that they have significant ethical implications. In a play on the famous law of increasing computing power, he coined "Moor's Law:"

> As technological revolutions increase their social impact, ethical problems increase. (p. 117)

Moor points to revolutionary technologies such as advanced information technology, genetic technology, nanotechnology, and neurotechnology, and notes that they have two salient features from an ethical viewpoint: malleability and convergence. By "malleability" he means they can be shaped to perform different tasks or to serve different functions. Genetic technology, for example, exhibits "life malleability" whereas nanotechnology is characterized by "material malleability." Moor breaks down the popular notion of convergence by noting that these technologies can serve as tools, components or models of one another. The implications of malleability and convergence are that:

> The ethical issues that we will confront will not only come in increasing numbers but will come packaged in terms of complex technology. Such ethical issues will require considerable effort to be understood as well as a considerable effort to formulate and justify good ethical policies. This will not be ethics as usual. People who both understand the technologies and are knowledgeable about ethics are in short supply just as the need is expanding. (p. 118)

Moor concludes that three improvements are needed in our approach to revolutionary technology: acknowledgement that "ethics is an ongoing and dynamic enterprise;" employing a multi-disciplinary approach that includes "better collaborations among ethicists, scientists, social scientists, and technologists;" and development of "more sophisticated ethical analysis." Unfortunately, he stops short of outlining what such changes in ethical analysis would entail.

While Moor seems more concerned about the impacts of revolutionary technology than the technologies themselves, bio- and nano-ethicist George Khushf (2006) is concerned that without better ethical reflection we will be denied the benefits of emerging technologies. And yet Khushf's conclusions are remarkably similar to Moor's.

Khushf focuses on two characteristics of the convergence of nanotechnology, biomedicine, information technology, and cognitive science (NBIC): their accelerating development rate and the oft-stated goal of NBIC convergence of enhancing human performance. Similar to Moor, he argues a new ethics is called for in the face of "radical" technologies:

> The more radical the possibilities associated with new technologies, the more radical the ethical challenge, not just to the current norms of human interaction but also to the form and character of ethical reflection itself. At each stage, a new accommodation between the older wisdom and new context is demanded. (p. 255)

The rapid advancement of the NBIC technologies, Khushf warns, threatens to overwhelm our ability to make ethical decisions:

> My point, however, is not simply that we can expect many ethical issues to arise out of NBIC convergence....We are already approaching a stage at which ethical issues are emerging, one upon another, at a rate that outstrips our capacity to think through and appropriately respond....On the immediate horizon arises a point at which the traditional way we have addressed ethical issues fails, because it does not and cannot keep up with the rate at which new challenges emerge. Faced with the prospect of increasingly accelerating, radically new technologies, we must completely reassess how ethical issues are addressed and how ethical debate informs broader public and legal policy. The promise of NBIC convergence thus poses an ethical challenge not just in the number, scope, and depth of issues that are raised but also in the very form that ethical reflection takes. (p. 258)

In particular, Khushf argues that the traditional method of *post hoc* ethical reflection (usually from external critics) needs to be replaced by ethical reflection that is part and parcel of the R&D process. This, he argues, will not only be beneficial from an ethical viewpoint but also vital for the success of the technologies:

> ...the rate-limiting step in the emergence of radically new integrative technologies will be sociocultural and ethical, not scientific or technological. If we cannot develop new processes for reliable modes of ethical reflection – and by this, I mean forms of ethical reflection that embody the interests, concerns, and modes of reasoning that currently come in as a secondary, external step – then we face a sociocultural barrier to the rate of acceleration in NBIC domains. This is something that should be addressed now, at the beginning, as a part of initial formation of the culture of NBIC convergence.... (p. 261)

Like Moor, Khushf concludes this will require greater collaboration between scientists and humanists, going so far as to suggest it will require nothing less than the bridging of C. P. Snow's two cultures (1993), i.e., the sciences and the humanities.

Also like Moor, Khushf falls short in outlining new forms of ethical analysis appropriate to the NBIC revolution, except to make an eloquent case for the promoters of NBIC convergence to continue to promote the goal of human enhancement while trying to find common ground with critics of the concept.

3.7 Conclusion

Promoters and critics of emerging technologies, as well as engineering ethicists, computing ethicists, and bioethicists, all seem to acknowledge the unique character of emerging technologies due to such factors as convergence, embeddedness, malleability, and human transcendence. Such factors would seem to suggest that the traditional "ethical concepts," noted by Shrader-Frechette at the beginning of this chapter, may require reexamination. Indeed, the promoters of emerging technologies seem willing to abandon traditional concepts such as the privileged role of human agency in moral decision making. The critics and ethicists, on the other hand, seem more concerned with altering the *process* of ethical deliberation both in terms of timeliness and participation.

Pathways to an ethical middle ground, where both traditional concepts *and* processes are critically examined in light of emerging technologies, have yet to be mapped. There has been, however, some encouraging movement in that direction, including the following:

Moral imagination, a concept widely applied in practical and professional ethics, highlights the need to consider moral dilemmas from different perspectives and the points of view of different stakeholders. Dealing effectively with such characteristics of emerging technologies as complexity and malleability will require that moral imagination be applied liberally. Berne (2005), for example, has shown how moral imagination can be applied in science fiction to highlight the ethical challenges posed by nanotechnology.

Preventive ethics, a concept developed in the healthcare field to encourage development of policies, including policies relating to new technologies, before ethical dilemmas arise, has been applied by Harris (1995) to the problem of technological disasters. By learning from past ethical failures and anticipating ethical problems before they occur (e.g., by establishing better procedures for internal dissent) Harris argues that many disasters arising from ethical failures can be avoided. Preventive ethics thus echoes the call of Moor, Khushf, and others for more proactive ethical deliberation as opposed to the more common post hoc method.

Scholars of Science and Technology Studies (STS) have for some time been arguing for an engineering ethics that goes beyond individual moral responsibility to consideration of the societal context of engineering and technology (Lynch and Kline 2000, Johnson and Wetmore 2009). Such analyses will become more critical as emerging technologies remake sociotechnical systems. One insight offered by STS is how to incorporate the public in resolving increasingly complex questions of technology policy (e.g., through consensus conferences). Johnson (see her chapter in this volume) has also built on STS concepts in promoting *anticipatory ethics* that involves "engagement with the ethical implications of a technology while the technology is still in the early stages of development, engagement that is targeted to influence what is developed."

Based on the new technique of real time technology assessment, Allenby and Sarewitz (2011) are advocating the need for *Real Time Macroethical Assessment* aimed at addressing the concerns of Moor, Khushf and others that ethical deliberations need to be conducted concurrent with the development of emerging technologies and responsive to the novel characteristics of such technologies

Whether these and other new approaches to ethics will be developed in time remains to be seen. What is clear, however, is that if we don't accept the challenge of crafting new processes and concepts for ethical deliberation, our machines may one day be making these decisions for us.

References

Allenby, B.R., and D. Sarewitz. 2011. The Techno-Human Condition. MIT Press.

Anon. 2000. The future of humanoid robots. *Discover – New York-* 21: 84–90.

Arkin, Ronald C. 2008. Governing lethal behavior: embedding ethics in a hybrid deliberative/reactive robot architecture. In *Proceedings of the 3rd ACM/IEEE international conference on human robot interaction,* 121–128. Amsterdam, The Netherlands: ACM.

Berne, Rosalyn W., and Joachim Schummer. 2005. Teaching societal and ethical implications of nanotechnology to engineering students through science fiction. *Bulletin of Science Technology Society* 25(6) (Dec. 1): 459–468.

Hall, J. Storrs. 2007. *Beyond AI: Creating the conscience of the machine.* Amherst, NY: Prometheus Books.

Harris, C.E. 1995. Explaining disasters: the case for preventive ethics. *Technology and Society Magazine, IEEE* 14 (2): 22–27.

Herkert, Joseph. 2001. Future directions in engineering ethics research: Microethics, macroethics and the role of professional societies. *Science and Engineering Ethics* 7 (3): 403–414.

Herkert, Joseph. 2000. Engineering ethics education in the USA: Content, Pedagogy, and curriculum. *European Journal of Engineering Education* 25 (4): 303–313.

Johnson, Deborah G., and Jameson M. Wetmore. 2009. *Technology and society: Building our sociotechnical future.* Inside technology. Cambridge, MA: The MIT Press.

Khushf, George. 2006. An ethic for enhancing human performance through integrative technologies. In *Managing nano-bio-info-cogno innovations* 255–278. http://dx.doi.org/10.1007/1-4020-4107-1_16.

Ladd, John. 1985. The quest for a code of professional ethics: an intellectual and moral confusion. In *Ethical issues in the use of computers*, 8–13. Belmont, CA: Wadsworth Publ. Co.

Lynch, William T., and Ronald Kline. 2000. Engineering practice and engineering ethics. *Science, Technology, and Human Values* 25 (2): 195–225.

Moor, James. 2005. Why we need better ethics for emerging technologies. *Ethics and Information Technology* 7 (3): 111–119.

Nordmann, Alfred. 2004. *Converging technologies: Shaping the future of European societies; report 2004.* EUR, 21357. Luxembourg: Office for Official Publ. of the Europ. Communities.

O'Connell, Brian M., and Joseph Herkert. 2004. Engineering ethics and computer ethics: Twins separated at birth? *Techné: Research in Philosophy and Technology* 8 (1): 36–56.

Pasulka, N. 2008. More than meets the eye. The morning news (May 2). Available at http://www.themorningnews.org/archives/in_hindsight/more_than_meets_the_eye.php.

Shrader-Frechette, Kristin. 1997. Technology and ethical issues. In *Technology and values,* eds. K.S. Shrader-Frechette, and Laura Westra, 25–30. Lanham, MD: Rowman & Littlefield Publishers.

Snow, C.P. 1993. *The two cultures*. London: Cambridge University Press.
Timmins, Peter. 2006. Privacy is dead – get over it. Centre for Policy Development (Apr. 19). Available at http://cpd.org.au/article/%2526%2523039%3Bprivacy-dead-%E2%80%94-get-over-it%2526%2523039%3B.
Wulf, William. 2003. Keynote Address. In National Academy of Engineering., National Academies (US), and National Academies Press (US), *Emerging technologies and ethical issues in engineering: papers from a workshop, 14–15 October*, 1–6. Washington, DC: National Academies Press.

Part II
Oversight Dynamics for Emerging Technologies

Chapter 4
Public Policy on the Technological Frontier

David Rejeski

It is difficult to pinpoint exactly when America moved from the geographic to the technological frontier. That moment may have been on July 12, 1893 when a young historian named Fredrick Jackson Turner declared that the country's westward expansion – a movement that had shaped the American psyche – was over. Turner's obituary for the frontier coincided with the Chicago World's Fair, a six-month love fest with architecture and technology that featured the first glimpse of what electricity might bring to American society, from lighting to motion pictures. Turner noted that, "In this advance, the frontier is the outer edge of the wave. . ..but as a field for the serious study. . .it has been neglected" (Turner 1893).

Today, the technological frontier remains a backwater to be experienced but seldom studied. Public policy makers operate daily on this frontier, but travel with little guidance and significant conceptual baggage. Like our forefathers on the geographical frontier, those on the technological frontier confront what Peter Bernstein has called the "wildness" – a world of change and uncertainty that confounds easy decisions, undermines predictions, and can often lead to embarrassing miscalculations by decision-makers. As Bernstein noted, "It is in [the] outliers and imperfections that the wildness lurks" (Bernstein 1996). Besides rampant uncertainty, the technological frontier shares one similarity with the old frontier – bad things can and do happen. Accidents are "normal" on the frontier, a point that Charles Perrow pointed out years ago (Perrow 1984). Despite the uncertainties, the frontier is where the expectations develop that shape business strategies, public opinion, and government actions over time (Bonini et al. 2006).

There are a host of issues that make navigating the technological frontier difficult for government entities including: *novelty* that undermines prediction, *cognitive biases* that blur our perceptions, *framing* that distorts emergent debates on public policies, *intractable problems* with too little funding to solve them, and a host of *known unknowns* that go unaddressed. One issue that has begun to attract more interest, and concern, is what some see as a growing mismatch between the rate of

D. Rejeski (✉)
Woodrow Wilson International Center for Scholars, Washington, DC, USA
e-mail: David.Rejeski@wilsoncenter.org

G.E. Marchant et al. (eds.), *The Growing Gap Between Emerging Technologies and Legal-Ethical Oversight*, The International Library of Ethics, Law and Technology 7,
DOI 10.1007/978-94-007-1356-7_4, © Springer Science+Business Media B.V. 2011

innovation in the public and private sectors. The basic argument is that this mismatch presents government with a quandary: Either speed up, which could lead to ill-considered actions or poorly conceived policies, or become irrelevant and incapable of impacting the dynamics of technological change (Popper 2003).

How serious this problem is depends on whether one believes there is an expanding gap in innovation rates – a tortoise-and-hare problem. There is evidence that the time it takes to introduce new technologies has been shrinking. Between 1990 and 1995, the time to develop and introduce US products fell from 35.5 to 23 months and the time needed to introduce high-tech products into the marketplace dropped from 18 months in 1993 to 10 months in 1998 (Griffin 1997; Tassey 1999). Taking a longer historical look, Yale university economist William Nordhaus has estimated that about 70 percent of all goods and services consumed in 1991 were different from those of a century ago (Nordhaus 2009). In the period from 1972 to 1987, the US government eliminated 50 industries from its standard industrial classification (SIC codes). In the decade following 1987, the government deleted 500 and added, or redefined, almost 1,000.

There is a tendency to evoke Moore's Law – Gordon Moore's 1965 prediction that the performance of integrated circuits would double every 18–24 months – as a metric of today's rapid innovation tempo. However, the distance between computer chips and actual computers is large and the gap is littered with failed startups and wasted capital. Bhaskar Chakravorti coined the term *Demi-Moore's Law* to indicate that technology's impact on the market moves at a rate only one half the speed predicted by Gordon Moore (Chakravorti 2003). As Clayton Christensen at Harvard Business School has noted, technologists have a habit of overestimating consumer demand and often project huge markets that never materialize. It's been jokingly said that computer scientists, looking at new markets, count 1, 2, 3,... a million (Seely Brown and Duguid 2000). Regardless of the absolute rate of change, the relative distance between private sector innovation and public sector response seems to be growing.

In one emerging area – nanotechnology – a growth of patents has yielded a correspondingly rise in products on the market with a 10–12 year lag between invention and market penetration. The number of manufacturer-identified, nano-based products on the market has risen from around 50 in 2005 to over 1,000 in August, 2009, and to 1,300 by the end of 2010. A linear regression model fitted to this trend data projected 1,700 products by 2013 (R2 = 0.996) (Project on Emerging Technologies 2011).

As nanotechnologies were introduced into the marketplace, a secondary lag occurred between the introduction and an understanding of any risks to human health and the environment – a lag that is likely to grow. A recent study on the potential costs and time required to assess the risks of just 190 nanomaterials now in production indicated a required investment of $249 million (assuming optimistic assumptions about hazards and streamlined testing techniques) to almost $1.2 billion to implement a more comprehensive battery of tests in line with a precautionary approach (this approach would require between 34 and 53 years to implement)

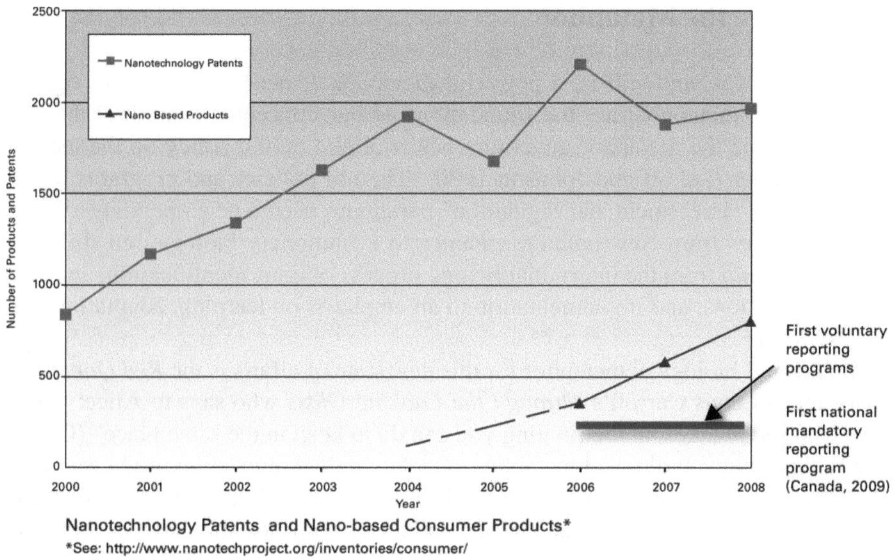

Nanotechnology patents (Chen et al. 2008) and Nano-based consumer products

(Choi et. al. 2009). Keep in mind that the risk assessment challenge is likely to increase in complexity and cost with second and third generation nanotechnology products and materials. A third lag then occurred between the recognition of risks and attempts to manage them, either through voluntary approaches or mandatory reporting requirements and regulations. A new comparative US-EU report calls for mandatory reporting for nanomaterials in commercial use but, to date, only the Canadian government has implemented this type of regulation (Breggin et al. 2009).

The shock of the new is compounded by what English historian David Edgerton called "the shock of the old" (Edgerton 2007). Once introduced, technologies tend to linger, often for decades. Our strategic arsenal still relies on the B-52 bomber (in service since 1955), machetes and small arms kill most people in wars, and our environmental policies still focus on technologies developed during the last industrial revolution, such as the internal combustion engine, steam-powered electricity generation, and bulk chemical synthesis. The organizational challenge is dealing with three types of technologies simultaneously: old technologies from the past, old technologies combined in new ways, and the truly new and novel. So the flood of emerging nanoscale materials, many with highly novel properties, comes on top of 80,000 chemicals already in commerce that we know very little about in terms of their risks to humans and the environment (Environmental Defense Fund 1997).[1]

[1]For decades, we have had inadequate human health risk data on most of the chemicals in commerce, less information on ecological risks, and virtually no data on synergetic effects and risks. In 1984, the National Academy of Sciences' National Research Council published a four-year study

4.1 Change the Metaphor

The frontier was, and still is, a powerful metaphor. If neuroscientists are right in asserting that metaphors are the foundations of our conceptual thinking, then we need to change the metaphor governing behavior and public policy on the technological frontier (Lakoff and Johnson 1980). The old policies and programs, based largely on an "assessment and regulation" paradigm, need a new operating system, one that moves from Newtonian mechanics to evolutionary biology and shifts the *modus operandi* from the interminably long process of issue identification, analysis, recommendations, and implementation to an emphasis on learning, adaptation, and co-evolution.[2]

One useful biological metaphor for this new state of affairs is the *Red Queen*, the character in Lewis Carroll's *Through the Looking Glass* who says to Alice: "Now, here, you see, it takes all the running you can do to keep in the same place" (Carroll 1872). Applying a biological metaphor to technological innovation might seem far-fetched, but the question is what we might learn from such an analogy.[3] As Stuart Kaufman once noted, "What can biology and technology possibly have in common? Perhaps nothing, perhaps a lot" (Kauffman 1995). Catching up with technological innovation is difficult and our governance institutions are handicapped by the existing approach to policy design – slow, expensive, and hard to maneuver in the face of change, uncertainty, and conditions of constant surprise. In this situation, metaphors matter because they serve as a means of structuring, and potentially changing, how we see, think, and act. Organizations viewed as machines, for instance, will operate very differently from organizations viewed by their members as brains or adaptive organisms (Morgan 1997).

One response to the Red Queen would be a shift from serial to parallel processing, or, to use an approach from the business world, a move towards concurrent engineering where product and process design run simultaneously, achieving time savings without sacrificing quality. Applying the Red Queen metaphor to public policy challenges on the technological frontier has three important implications for the behavior of organizations:

- First, co-evolution is the only operable strategy. As John Seely Brown, the former head of the Xerox Palo Alto Research Center (PARC), once observed "The

---

and found that 78 percent of the chemicals in highest-volume commercial use had not had even "minimal" toxicity testing. Today, there has been little improvement (National Research Council 1984). That is the problem we have inherited which will combine with new risks from emerging technologies.

[2]A 1972 analysis of technology assessment revealed that most assessments cost between $800,000 and $2 million and took 16–18 months to complete – not much has changed since then with assessments today by organizations such as the National Academy of Sciences taking up to two years and often costing at least $1.5 million (Coates 1972).

[3]In nature, the high-tempo Red Queen may not drive evolution on a continuing basis, but be balanced by stable strategies in which various actors are better off not changing their strategies.

future is not invented; it is co-evolved with a wide class of players." The players in the policy system become part of a diverse, complex, and dynamic innovation ecosystem, not isolated observers sitting on some external perch. The goal is to prevent risks, not just study them; to encourage innovation, not just write about it; and to accelerate the introduction of sustainable technologies into the marketplace, not to hinder it.

- Second, time matters. Understanding the pace of change of the actors in the innovation system will define strategies (for instance, shaping or adapting, and impact actions, such as placing big bets or creating options and no-regrets moves) and determine the nature and ultimate outcomes of co-evolution (Courtney et al. 1997). This sense of time and timing depends on a high degree of situational awareness or what some term "mindfulness" of the environment, constraints, opportunities, and expectations (Weick and Sutcliffe 2001). One key piece of information is an understanding of the decision cycle of key actors in the ecosystem – from industry to the Congress – and being able to gain influence or competitive advantage by getting inside that cycle.[4]
- Finally, change/learn or die. One of the most important implications of the Red Queen metaphor is that previous behaviors and adaptations do not guarantee continued survival in the face of future challenges (Hoffman 1991). One has to effectively learn from the past, but adaptive learning on the fly is also critical and that implies continual experimentation with innovative methods and organizational structures.

Imagine a new set of functions designed to operate dynamically inside the innovation system in a parallel processing mode that focus on co-evolution and rapid learning. This list is not exhaustive, but exemplary, and designed to form the basis of an experimental and empowering niche that could support a broader transition to new policies and organizational strategies (Rotmans and Loorbach 2009).

4.2 Embed an Early Warning System

Without early warning, early action is difficult and a reactive response is almost preordained. Proponents of reflexive or anticipatory governance have raised the issue of early warning but little action has been taken on the part of government to institutionalize the function (Guston and Sarawitz 2002).

Here is one example of an early warning failure on the technological frontier. Concerns about possible inhalation risks of carbon nanotubes first appeared in a letter by industrial hygienist Gerald Coles written to *Nature* magazine in

[4]One useful model for understanding decision loops was developed by former Air Force fighter pilot, John Boyd, and is know at OODA (which stands for observe, orient, decide, and act). Some of John Boyd's key writing can be found at: http://www.d-n-i.net/dni/john-r-boyd.

1992.[5] In 1998, science journalist Robert Service wrote an article in *Science* magazine entitled: "Nanotubes: The Next Asbestos?" again raising concerns, which were downplayed by a number of nanoscientists, including Nobel prize winner Richard Smalley (Service 1998). As was recently noted, Smalley "...did not want to draw attention to the hypothetical dangers of nanotechnology in case it would undermine support for the field in the early days" (Toumey 2009). Fast-forward another decade and more evidence has accumulated that carbon nanotubes can cause asbestos-like pathogenicity in the lung and actually pass directly through the lung lining (Poland et al. 2008; Sanderson 2009). Recently, the Environmental Protection Agency declared it would finally enforce pre-manufacturing reviews for carbon nanotubes, declaring that carbon nanotubes "are not necessarily identical to graphite or other allotropes of carbon."[6] This represents a minimal gap of over 15 years between early warning and regulatory action. During this time little funding was invested by government to resolve initial concerns and risks were in many cases actively downplayed by researchers and developers. This early warning was possible based on a structural analogy to a known, and highly toxic material – asbestos. Although the hallmark of innovation in areas like nanotechnology and synthetic biology is their ability to destroy analogy, to create novel materials and organisms with no historical referents that can guide prediction, there are nevertheless historical precedents and lessons that can provide valuable warning signals.[7]

In a Red Queen world, warning moves along with the science, it does not come after the fact, especially after materials and products have already been introduced into commerce. One approach would be to establish in all oversight agencies – the Environmental Protection Agency, Food and Drug Administration, Department of Agriculture, and Consumer Product Safety Commission – an early warning officer (EWO) with associated support staff (3–4 full-time equivalents). The EWO would report directly to the head of the agency and provide once-a-month briefings that focused not just on threats, but on opportunities to leverage emerging technologies to improve the agency's mission. Early Warning Officers from multiple agencies could also meet to exchange information on a regular basis and build a larger network that encompassed state, local, and international members. This type of strategic reconnaissance is fairly common in the business and intelligence sectors, so those models could be easily adapted to oversight organizations.

---

[5]"Sir – Attractive though they are, the technical properties of ultra-thin man-made fibres pointed out by Paul Calvert (Nature 357 365; 1992) should not hide the potential – at least for those fibres resistant to biological degradation in vivo – for related occupational risks to workers."

[6]See: "EPA to Enforce Premanufacturing Reviews for Carbon Nanotubes Beginning March 1. Reported at: http://www.merid.org/NDN/more.php?id=1728. And: Toxic Substances Control Act Inventory Status of Carbon Nanotubes, 73 Fed. Red. 64946 (31 Oct 2008).

[7]A calculation done at Rice University indicated that by simply modifying a number of variables of the 20 major types of single walled nanotubes – variables involving manufacturing process, tube lengths, methods of purification, and possible surface coatings – over 50,000 possible variants of this one nanomaterial were possible (Kulinowski 2008). Which ones pose risks? Given the large and growing uncertainty around emerging risks, significant effort and funding needs to be focused on techniques like tiered screening and high throughput testing.

4.3 Track the Known Unknowns

When Wired Magazine called the EPA, FDA, and US Patent Office to ask about regulatory approaches to the emerging area of synthetic biology, the agency people had to ask what synthetic biology was (Keim 2007). As a new scientific field emerges there is far more that we don't know about possible risks, unintended consequences, and governance options than we know. As Robert Proctor, an historian of science at Stanford, once noted "[It] is remarkable how little we know about ignorance" (Proctor and Schiebinger 2008). Ralph Gomory, the former president of the Sloan Foundation, once wrote a provocative essay on the unknown and unknowable, noting that "We are all taught what is known, but we rarely learn about what is not known, and we almost never learn about the unknowable. That bias can lead to misconceptions about the world around us" (Gomory 1995). One approach would be to develop an open-source tool that provided an evolving list of *known unknowns* for an emerging area of science and technology. As empirical evidence was gathered, issues could be modified, taken off the list, or new areas of inquiry added. For instance, in the area of synthetic biology, one unknown at the moment is: How to best assess the risks of novel organisms with little or no natural precedents? An evolving list of known unknowns (possibly maintained on a Wiki) would also constitute a de facto risk research agenda that could be addressed by national and international funders. Finally, it may reduce the potential for surprises, allowing policymakers the opportunity to consider various scenarios *before* they occur.

This exercise does not address the unknown unknowns or unknowables, but a continual focus on unknowns may force policymakers and researchers to begin to discriminate more carefully between various classes of unknowns and pay attention to building more flexible and adaptive organizations which can respond to surprises or events that occur beyond the realm of normal expectations (so-called Black Swans) (Talib 2007).

4.4 Focus on Bad Practices

It is common for those operating on the technological frontier to focus on best practices, often singling out particular companies and operations for awards. This is important but, paradoxically, one of the most important things to do when confronted with high degrees of technological uncertainty is to focus on the bad practices. Every single day vigilant and intelligent people recognize errors around them and can often come up with ingenious ways to correct problems. Taken one at a time, these bad practices seldom lead to a disaster, if recognized early and addressed. The challenge is to develop ways for "error correcting knowledge" to be collected, managed effectively, and channeled into solutions. One model for this is the Aviation Safety Reporting System (ASRS), which collects and analyzes voluntarily submitted reports from pilots, air traffic controllers, and others involving safety risks and incidents.[8] Operated by NASA for the aviation industry, ASRS

[8] See: http://asrs.arc.nasa.gov/.

is described as confidential, voluntary, and non-punitive. The reports are used to remedy problems, better understand emerging safety issues, and generally educate people in the aviation industry about safety. A similar system in the UK, called CHIRP, is designed to promote greater safety in both the aviation and maritime industries and is run by a charitable trust.

One option is to create a Safety Reporting System for emerging areas of science and technology where concerned people working in laboratories, companies, or elsewhere can anonymously share safety issues and concerns. The purpose is not "finger pointing" but encouraging proactive learning before something goes really wrong. Information could be used to design educational materials, better structure technical assistance programs, and provide a heads-up on a host of emerging safety issues.

If these systems fail, there is a final backstop before some disaster hits – internal audits by inspector generals and, finally, whistleblowers.[9] Whistleblowers are the ones who watch the watchmen, often risking their careers to rise above their bureaucratic brethren. They are the antidote to group think, to the perceived invulnerability of the organization, the rationalizations, and insulation from outside opinion (Sonnenfeld 2005). The price is high. One half to two thirds of all whistleblowers lose their jobs (Alford 2001). Despite recent efforts to shore up whistleblower protections (in the Consumer Product Safety Improvement Act and the Whistleblower Protection Enhancement Act) one group remains largely unprotected – government employees. Strong whistleblower protection, especially in our regulatory agencies, is absolutely necessary as scientific innovation moves rapidly forward.

4.5 Get the Right People to the Frontier

One way to provide oversight and governance of science is to have the scientists and engineers provide it themselves – an approach that has been put forth in the areas of nanotechnology and synthetic biology. Whatever historical precedents existed for this type of reflective self-governance are long gone. As Steven Shapin has pointed out in his recent exploration of the moral history of science, there are no real grounds today "to expect expertise in the natural order to translate to virtue in the moral order" (Shapin 2009). Recent survey work with university-based nanoscientists has indicated that researchers working on new technologies tend to view their work as not producing any "new" or substantial risks, while those scientists downstream

[9]Recently, the Department of Energy (DOE) issued a comprehensive memo covering the "Safe Handling of Unbounded Engineered Nanoparticles" in DOE facilities. What preceded this directive was a scathing report by DOE's Inspector General that indicated that 11 out of 12 DOE labs did not perform medical surveillance of individuals working with nanoscale materials and 9 or the 12 labs had not initiated monitoring for exposure rates in the workplace. The report concluded that DOE should "adopt a proactive approach to ensuring that its laboratories follow best practices in conducting nanoscale-related work" (Department of Energy 2008).

of development often feel the exact opposite (Powell 2007). In addition, computer simulations of diverse problem solvers indicate that specialists often become trapped in suboptimal solutions to complex problems such as risk assessment (Hong and Page 2004).

Normally, people entering a frontier space are trained. Astronauts receive an average training of two years and brain surgeons undertake a six-year residency. This training promotes a professionalism that includes ethical components. But what about scientists and engineers operating on a technological frontier? A survey of over 250 accredited engineering programs in 1996–1997 found that only 1 in 5 offered students any significant exposure to ethics (Stephan 1999). Bill Wulf, who headed the National Academy of Engineering (NAE), said recently that "The complexity of newly engineered systems coupled with their potential impact on lives, the environment, etc., raise a set of ethical issues that engineers have not been thinking about," and the NAE recently established a new Center for Engineering, Ethics, and Society to meet the challenges (Dean 2008).

As a backup for training approaches, one could also embed social scientists in the research enterprise, an approach some have called "lab-scale intervention" designed to enhance direct interaction between different social and natural science disciplines during the research phase (Schuurbiers and Fisher 2009). This approach is undoubtedly better than having scientists operate with little or no feedback on the social and ethical impacts of their research. But one problem is that the same organization (such as the National Science Foundation) often funds both the researchers and the social scientists with the same grant, creating a co-dependency situation that certainly has the potential to compromise the social oversight function. Adding a few bioethicists or nanoethicists to the scientific mix to watch for missteps still leaves open the question: "quis custodiet custodes ipsos?" (who guards the guardians themselves?) or the more modern version: "Who watches the watchmen?" (Moore and Gibbons 1987).

4.6 Develop and Implement a Learning Strategy

A recent article on technological innovation made the point that, "in an era of complex technologies, and that will surely be the dominant characteristic of the early part of the twenty-first century, public policy will need to facilitate learning and be ever more adaptable" (Rycroft 2006). The more experiments one can run, the more hypotheses one can test, the faster the rate of learning. It sounds paradoxical but in terms of learning and innovation, "Whoever makes the most mistakes wins" (Farson and Keyes 2002). Over the last few decades, the economics of experimentation have dropped dramatically in the private sector because of advances in computation and rapid prototyping systems as well as an increasing focus on testing new organizational and leadership paradigms.

Unfortunately we seldom crash test public policies, but instead wait for them to crash. When EPA launched a voluntary program to gather information on nanomaterials, a number of experts, drawing on years of research on voluntary agreements, warned that the program would be ineffective without stronger incentives for industry participation and the backup of mandatory measures. The EPA program took three years to implement, during which time a similar program, launched in the UK, failed moving to yield the needed information on emerging nanoscale materials. EPA persisted forward – slowly. Not surprisingly, critics at the end of this tedious experiment noted that, "With hundreds of nano products already on the shelves, EPA has squandered precious time while it slowly developed and pursued a program that informed stakeholders cautioned would not yield what was needed" (NanoWerk News 2009). EPA pursued an internally focused, serial processing strategy, not a co-evolutionary, time-sensitive approach.

It is not clear that the agency had, or has, a clear learning strategy, one that can mitigate the probability of future errors by either learning from past efforts (where applicable analogies hold), from parallel efforts by other credible actors, or from thinking smarter about the future (Garvin 2000). In this regard, it is important to remember that, "experiments that result in failure are not failed experiments" (Thomke 2003). The organizational pathologies that undermine learning in organizations are well documented and include: (1) insulation from outside expert opinions, (2) fixation on single paths, (3) no contingency planning, (4) an illusion of invulnerability, (5) collective rationalization, (6) the denigration of outsiders, and (7) a coercive pressure on dissenters (Sonnenfeld 2005). Prevalent maxims are also well researched and well known: learn from failure, refuse to simplify reality, commit to resilience and flexibility, don't overplan (keep options open), and hire generalists (they'll thrive longer in complex ecosystems) (Weick and Sutcliffe 2001). Given the large and looming retirement bulge in many US regulatory agencies, like EPA, we have an opportunity to restructure the workforce in new ways that could address learning issues.

4.7 Conclusion

In a recent McKinsey survey on what factors contribute most to the accelerating pace of change in the global business environment, the top response was "innovation in products, services, and business models (Becker and Freeman 2006). The actual rate of technological change came in sixth place. The point is that it is not just technology, but technology's impacts on organizational strategy and ways of doing business, that cause an acceleration in innovation rates (for instance, the impact of high-speed computing on the entertainment or automobile industries). Charles Fine used an overarching approach in defining what he called organizational "clockspeed" – an evolutionary lifecycle defined by the rate a business introduces new products, processes, and organizational structures (Fine 1998).

Ultimately, this means that "pacing" governance to technological change will require focusing on the entire operating environment rather than just the technological components. This larger environment includes organizational structure, leadership, and securing talent as a strategic asset.[10] Counting bits per minutes or product introductions obscures the nature of the challenge that governments face. Viewed through a purely technological lens, the gap in innovation rates seems inevitable and insurmountable. Recognized and addressed as a "learning" problem provides some hope.

That does not mean the change in the public sector will be easy or fast. Organizations – in both the public and private sectors – often end up in what has been called a "competency trap" applying outmoded skills to emerging challenges (Levitt and Marsh 1988). By the time they catch up, competitive forces have created the next competency trap vis-à-vis a new set of actors and technological realities. In this situation, absolute speed becomes less critical than adaptive strategies because, as in evolution, competition and learning reinforce each other (Van Valen 1973). If we view biological and business evolution as complex adaptive systems, then the challenge for governments is to join the co-evolving system (Beinhocker 1999). That means turning a cognitive corner and seeing this rapid technological change as a learning and co-evolution challenge rather than just trying to run faster on the technological treadmill. In the end, disruptive innovation will require the application of disruptive intelligence in our public sector (McGregor 2005).

References

Alford, C.F. 2001. *Whistleblowers: Broken lives and organizational power.* Ithaca, NY: Cornell University Press.

Becker, W.M., and V.M. Freeman. 2006. Going from global trends to corporate strategy. *McKinsey Quarterly* 3: 17–27.

Beinhocker, E.D. 1999. Robust adaptive strategies. *Sloan Management Review* (Spring 1999): 95–106.

Bernstein, P. 1996. *Against the Gods: The remarkable story of risk.* New York, NY: Wiley.

Bonini, S. et al. 2006. When social issues become strategic. *McKinsey Quarterly* 2: 20–32.

Breggin, L. et al. 2009. *Securing the promise of nanotechnologies: Towards transatlantic regulatory cooperation.* London, England: Chatham House.

Carroll, L. 1872. *Through the looking glass and what Alice found there.* London: Macmillan.

Chakravorti, B. (2003). *The slow pace of fast change: Brining innovations to the market in a connected world.* Boston, MA: Harvard Business School Press.

Chen, H. et al. 2008. Trends in nanotechnology patents. *Nature* 3 (March): 123–125.

Choi, J. et al. 2009. The impact of toxicity testing costs on nanomaterial regulation. *Environmental Science and Technology* 43 (9): 3030–3034.

Coates, V. 1972. *Technology and public policy: The process of technology assessment in the federal government.* Washington, DC: George Washington University Program of Policy Studies in Science and Technology.

Courtney, H.G. et al. 1997. Strategy under uncertainty. *Harvard Business Review* 75 (6): 67–79, Nov.—Dec.

[10]Two global surveys by McKinsey, one in 2006 and one in 2007, indicated that finding talented people is likely to be the single most important preoccupation for managers for the next decade and that far greater competition for talent can be expected (Guthridge et al. 2008).

Dean, Cordelia 2008. Handle with care. *The New York Times*. See: http://www.nytimes.com/2008/08/12/science/12ethics.html?ref=technology&pagewanted=print. 12 Aug.

Department of Energy. 2008. *Audit report: Nanoscale materials safety at the department's laboratories*. DOE Office of Inspector General, DOE/IG-0788.

Edgerton, D. 2007. *The shock of the old: Technology and global history since 1900*. Oxford University Press.

Environmental Defense Fund. 1997. *Toxic ignorance: The continuing absence of basic health testing for top-selling chemicals in the United States*. Washington, DC: Environmental Defense Fund.

Farson, R., and R. Keyes. 2002. *Whoever makes the most mistakes wins: The paradox of innovation*. New York, NY: Free Press.

Fine, C. 1998. *Clockspeed: Winning industry control in the age of temporary advantage*. New York, NY: Basic Books.

Garvin, D. 2000. *Learning in action: A guide to putting the learning organization to work*. Boston, MA: Harvard Business School Press.

Gomory, R. 1995. An essay on the known, the unknown and the unknowable. *Scientific American* 272: p. 120, June.

Griffin, A. 1997. PDMA research on new product development practices: Updating trends and benchmarking best practices. *Journal of Product Innovation and Management* 14 (6): 429–458.

Guston, D., and D. Sarawitz. 2002. Real-time technology assessment. *Technology in Society* 23 (4): 93–109.

Guthridge, M. et al. 2008. Making talent a strategic priority. *McKinsey Quarterly* 1: 48–59.

Hoffman, A. 1991. Testing the Red Queen hypothesis. *Journal of Evolutionary Biology* 4: 1–7.

Hong, L., and S. Page. 2004. Groups of diverse problem solvers can Outperform groups of high-ability problem solvers. *Proceedings of the National Academy of Sciences* 101 (46): 16385–16389.

Kauffman, S. 1995. Escaping the Red Queen effect. *The McKinsey Quarterly* 1: 119–128.

Keim, B. 2007. Will synthetic biology catch government by surprise? *Wired Science* 15 (7): See: http://www.wired.com/wiredscience/tag/synthetic-biology/page/4/, 5 July.

Kulinowski, K. 2008. Towards predicting nano-biointeractions: an international assessment of research needs for nanotechnology environment, health and safety. Available at: http://www.wilsoncenter.org/ondemand/index.cfm?fuseaction=home.play&mediaid=A1838026-0268-AD3E-AD56DCC4E2B1C977.

Lakoff, G., and M. Johnson. 1980. *Metaphors we live by*. Chicago, IL: University of Chicago Press.

Levitt, B., and J.G. Marsh. 1988. Organizational learning. *Annual Review of Sociology* 14:319–340.

McGregor, J. 2005. Gospels of failure. *Fast company*, Feb.

Moore, A., and D. Gibbons. 1987. *Watchmen*. New York, NY: DC Comics.

Morgan, G. 1997. *Images of organizations*. Thousand Oaks, CA: Sage Publications.

NanoWerk News. 2009. EPA's Voluntary Approach Captures Only A Thin Slice of Nanomaterials in the US. At: http://www.nanowerk.com/news/newsid=8862.php. 13 Jan.

National Research Council. 1984. *Toxicity testing*. Washington, DC: National Academy Press.

Nordhaus, W. 2009. The entrepreneurial society. *The Economist* 390 (8622): p. 20, 14–20 Mar.

Perrow, C. 1984. *Normal accidents: Living with high-risk technologies*. New York, NY: Basic Books.

Poland, C.A. et al. 2008. Carbon nanotubes introduced into the abdominal cavity of mice show asbestos-like pathogenicity in a pilot study. *Nature Nanotechnology* 3:423–428.

Popper, S.W. 2003. Technological change and the challenges for 21st century governance. In: *AAAS science and technology policy yearbook*, eds. (A). Teich, and S. Nelson. Washington, DC: AAAS.

Powell, M.C. 2007. New risk or old risk, high risk or no risk? How scientists' standpoints shape their nanotechnology risk frames. *Health, Risk & Society* 9 (2) (June):173–190.

Proctor, R., and L. Schiebinger eds. 2008. *Agnotology: The making and unmaking of ignorance*. Stanford, CA: Stanford University Press.

Project on Emerging Nanotechnologies. 2009. Analysis available at: http://www.nanotechproject.org/inventories/consumer/analysis_draft/.

Rotmans, J., and D. Loorbach. 2009. Complexity and transition management. *Journal of Industrial Ecology* 13 (2):184–196.

Rycroft, R. 2006. Time and technological innovation: Implications for public policy. *Technology in Society* 28: 281–301.

Sanderson, K. 2009. Migrating nanotubes add to asbestos concern. *Nature Online*. See: http://www.nature.com/news/2009/090331/full/news.2009.217.html, 31 Mar.

Schuurbiers, D., and E. Fisher. 2009. Lab-scale intervention. *European Molecular Biology Organization Report* 10 (5): 424–427.

Seely Brown, J., and P. Duguid 2000. *The social life of information.* Boston, MA: Harvard Business School Press.

Service, R. (1998). Nanotubes: The next asbestos? *Science* 281 (5379) (Aug. 14): 941.

Shapin, S. 2009. *The scientific life: A moral history of a late modern vocation.* Chicago, IL: University of Chicago Press.

Sonnenfeld, J. 2005. Why it's so hard to blow the whistle. *Yale Alumni Magazine*. See: http://www.yalealumnimagazine.com/issues/2005_03/forum.html, March/April.

Stephan, K.D. 1999. A survey of ethics-related instruction in US engineering programs *Journal of Engineering Education* 88 (Oct.): 459–464.

Talib, N. 2007. *The black swan: The impact of the highly improbable.* New York, NY: Random House.

Tassey, G. 1999. *R&D trends in the US economy: Strategies and policy implications.* Washington, DC: US Department of Commerce, Planning Report 99–62.

Thomke, S.H. 2003. *Experimentation matters: Unlocking the potential of new technologies for innovation.* Cambridge, MA: Harvard Business School Press.

Toumey, C. 2009. Science from the inside *Nature Nanotechnology* 4: 537–538.

Turner, F.J. 1893 The significance of the frontier in American history. *Report of the American Historical Association for 1893* 199–227.

Van Valen, L. 1973. A new evolutionary law. *Evolutionary Theory* 1: 1–30.

Weick, K.E., and Sutcliffe, K.M. 2001. *Managing the unexpected.* San Francisco, CA: Jossey-Bass.

Chapter 5
Software Agents, Anticipatory Ethics, and Accountability

Deborah G. Johnson

Classification does indeed have its consequences – perceived as real, it has real effects.
– Bowker and Star (1999)

5.1 Introduction

An important current trend in software development is to produce complex systems designed to operate independently from the humans who design and deploy them. Often referred to as *software agents*, at least some of these systems are able to learn and make second order decisions that effectively reprogram how they operate. Software agents may be deployed to perform fairly simple transactions (such as purchasing a product) or to perform extremely complicated operations involving sophisticated decision making (such as the software onboard the Earth Observation satellite that decides which events on earth the satellite should monitor (Chien et al. 2005; Noorman 2008)). Software agents are not distinctive insofar as they operate separately in space and time from those who deploy them, but rather because of the kinds of decisions they make and their capacity to learn and behave proactively often through processes that are not transparent to designers and users. Generally, this technology is understood to be an extension or species of what was earlier thought of as artificial intelligence.

For some the characterization of software as *agents* seems a fairly straightforward matter of describing software that performs tasks on behalf of humans:

> The Intelligent Software Agents Lab at Carnegie Mellon University's Robotics Institute envisions a world in which autonomous, intelligent software programs, known as software agents, undertake many of the operations performed by human users of the World Wide Web, as well as a multitude of other tasks. (http://www.cs.cmu.edu/~softagents/intro.htm)

D.G. Johnson (✉)
University of Virginia, Charlottesville, VA, USA
e-mail: dgj7p@virginia.edu

G.E. Marchant et al. (eds.), *The Growing Gap Between Emerging Technologies and Legal-Ethical Oversight*, The International Library of Ethics, Law and Technology 7,
DOI 10.1007/978-94-007-1356-7_5,

For others, the autonomous aspect of software agents is what is significant because it means that they can operate in open-ended situations. The aim of the annual AGENTS: International Conference on Autonomous Agents is explained in the ACM Portal:

> Autonomous agents are software and robotic entities that are capable of independent action in open, unpredictable environments. Agents are currently being applied in domains as diverse as computer games and interactive cinema, information retrieval and filtering, user interface design, electronic commerce, autonomous vehicles and spacecraft, and industrial process control. (http://portal.acm.org/browse_dl.cfm?linked=1&part=series&idx=SERIES134&coll=portal&dl=ACM)

For some the nature of the software justifies a category of autonomous cognitive agents in which software and robots are apparently together with humans, though distinguished as artificial:

> Autonomous cognitive agents, whether natural or artificial, are information processing entities that make decisions, recognize patterns, gather information and perform actions. The concept of autonomy refers to the ability to use experience to determine action. This includes being able to adapt behavior in order to pursue goals under changing circumstances. Artificial agents can take a range of forms from software agents to anthropomorphic robots. (Lee and Lacey 2003)

The metaphor of software as autonomous agents facilitates the use of other concepts, such as negotiation, that are helpful in describing the behavior and operation of software:

> Automated negotiation is a powerful (and sometimes essential) means for allocating resources among self-interested autonomous software agents. A key problem in building negotiating agents is the design of the negotiation strategy, which is used by an agent to decide its negotiation behavior. (Rahwan et al. 2007)

To think of software as an agent and describe it as autonomous is, as already indicated, to use a metaphor but why this metaphor? What role or function does the metaphor play in the discourse and the design of software agent technology? Metaphors are important in shaping human understanding, though they can be dangerous when they lead to false presumptions or hide key features of the thing being explained. How does the autonomous agent metaphor work in the case of software? Obviously it provides a way of thinking about software, but what is at stake for those who use the metaphor? Does it lead to false presumptions? And what does it hide?

Importantly, use of the metaphor connects software agents to a number of fictional entities described in popular literature and media. Whether we take the monster in Mary Shelly's *Frankenstein*, the computer, Hal, in the film *2001* or the robots in the more recent film, *I Robot*, connections to the discourse of software agents are obvious. Indeed, the popular literature seems to express a mixture of fascination with and fear of these technologically-created, human-like beings as does the discussion of software agents of today. The possibility of humans losing control of human-made entities was the major thrust of Bill Joy's "Why the Future Doesn't

Need Us" (2000). When it first appeared, Joy's piece unsettled many scientists and intellectuals because an insider – a leader in the science and engineering community – was expressing concern about, and even reluctance about going forward with combining genetics, nanotechnology, and robotics. The combination, Joy warned, might be so powerful as to produce entities that would take over the world and render humans irrelevant.

As will be discussed further in the next section, the ideas that circulate during the early stages of development of a new technology influence the construction of meaning as well as the material design of the technology. Ideas contribute to the delineation of a technology as something in particular. New technologies can challenge deeply rooted beliefs about what it means to be human; they can challenge the distinction between what is human and what is artificial; and raise daunting normative questions about how the human-technology relationship should be constituted. The discourses around new technologies both *express* these deep human concerns and *shape* the developing technology.

This chapter begins with the premise that the characterization of software as *autonomous agents* is influencing the meaning and material design of the technology; that is, use of these terms and the metaphor is having an effect on what will eventually be developed and how it will be understood. The chapter takes as its starting place the observation that using the metaphor of autonomous agents may be setting the scene for a collision with moral and legal notions and practices of accountability. If software agents behave autonomously, in ways not fully understood by their human designers or users, who will be accountable when something goes wrong? How will accountability be handled when the behavior of software agents leads to harmful consequences?

To comprehend the potential for collision, consider a worst case scenario. Suppose: (1) very powerful software agents are put to use in activities that have powerful consequences for human well-being; (2) no human beings are able to understand what these powerful agents have learned and how they make many of the decisions they make since they act autonomously; (3) a software agent's behavior is the major factor leading to a catastrophic event (e.g., an industrial accident, launch of a nuclear weapon, a major electricity shut down); (4) victims of the event and the public call for an explanation, that is, they demand that someone be held accountable for the event and the harm done; and (5) victims and the public are told that it was the behavior of the software agent that led to the event, that no human can understand why the agent did what it did and, therefore, no human beings are responsible for the behavior of the software agent.

Whether this scenario ever occurs, software agents are being designed to perform tasks, to learn as they operate, and to change their decision-making strategies for achieving designated tasks as they learn. When the behavior of a software agent results in harm to humans, issues of accountability are likely to arise. Yet the conceptualization of software agents seems to be setting the scene for a deflection of responsibility, at least *human* responsibility for the behavior of software.

5.2 Making Room for Anticipatory Ethics

Given the potential for collision between the development of software agents and prevailing moral and legal notions of accountability, software agent technology seems an ideal case for anticipatory ethics. *Anticipatory ethics* refers here to: (1) engagement with the ethical implications of a technology while the technology is still in the earliest stages of development; and (2) engagement that is targeted to influence the development of the technology. Software agent technology is in the early stages of development; thus, it offers the possibility of anticipating accountability issues now, rather than waiting until the technology is well developed and an untoward event occurs. The challenge is to see whether issues of accountability can somehow be taken into account and incorporated in the design and early thinking about this technology, with an eye to avoiding collision.

Since anticipatory ethics is a new approach to addressing ethical issues related to technology, some background will be helpful in understanding the endeavor. Currently, the most visible and well-funded attempt at anticipatory ethics in the US is focused on nanotechnology. A major impetus for this work has been a government mandate (and the availability of resources) to examine the social and ethical implications of nanotechnology. The twenty-first century Nanotechnology Research and Development Act (Public Law 108-193 passed in 2003) specified that the National Nanotechnology Infrastructure (NNI) include activities that ensure that "ethical, legal, environmental, and other appropriate societal concerns... are considered during the development of nanotechnology." A number of scholars have risen to the task and a growing literature on the social implications of nanotechnology and nanoethics has developed. (See, for example, the journal *Nanoethics* and most recently the *Yearbook of Nanotechnology in Society* (Fisher et al. 2008)). Since nanotechnology is still very much a technology (technologies) in-the-making, work in this area falls into the category of anticipatory ethics.

Why nanotechnology? Why now? Why address the social and ethical implications of a technology before it is ready for use? At least part of the explanation has to do with a shift in understanding of the processes of technological development. Until recently (i.e., the last several decades) and with a few exceptions, ethics scholars paid little or no attention to technology. Technology was considered irrelevant to ethics both because ethics was understood to be about human behavior and because technology was thought to be neutral – values lay in how humans used technology. Interest in technology began to develop in the last half of the twentieth century. Some would say it was the powerful social effects of an array of modern technologies including the atomic bomb, industrial chemicals, computers, and genetic engineering. Whatever the underlying causes, concerns about technology coincided with the movement in practical ethics. The attention of ethicists slowly turned to technology, especially computers and information technology.

Initially ethicists adopted a framework in which technology was understood to be, primarily, the outcome of the work of scientists and engineers (Johnson and Wetmore 2008). In this framework, whether we distinguish scientists and engineers and whether they work in the ivory tower, in corporations, or in government, the presumption is that technologies develop somewhat separately from society, and, when completed, are delivered to society. Society can then chose whether to adopt a delivered technology, and, if adopted, a technology may then have *social impacts*.

The task of ethicists is, in this framework, to examine the social impacts and to note and address how the introduction and adoption of a technology creates ethical issues and affects important social values.[1] The social impact framework presumes that scientists and engineers work in relative isolation, doing what nature dictates, and that the resulting technologies are neutral. Values come into play when humans decide whether or not and how to take up what has been delivered. Working within this framework, ethicists do not ask about, let alone examine, the social forces, the institutional actors, the interests, or the values that have directed attention and resources to a particular technological endeavor; nor do they examine the factors that determined the design features of a new technology. Engineers are understood to be *applying* science, and since nature dictates science, engineers are constrained by nature.

In this framework it may look like nature necessitates the kind of technology that is produced, i.e., the kind of airplanes, medical devices, and power plants that are "delivered to society." And, if nature dictates the character of technologies, then there is little room for ethics or values to come into play. The only role for ethicists (or consumers and users for that matter) is to decide whether or not, and how, to use the technology that scientists and engineers deliver. The primary role for ethics, in this framework, is reactive. Ethicists can critique what is delivered; for example, they can show how surveillance technologies violate privacy. They can call for modifications in design; for example they can call for wider sidewalks and ramps next to stairways to ensure access by those confined to wheelchairs. Or ethicists can analyze social practices involving technology; for example, ethicists have analyzed the fairness of various procedures for distributing scarce medical resources. In this mode of operation, it is not surprising that ethicists may be accused of being *anti-technology* for in their reactive role, they are more likely to notice technologies that disrupt or threaten moral practices or values than to notice those that fit neatly in or enhance prevailing moral practices and values.

To be sure, ethicists working in the social impact framework have made important contributions; the fields of biomedical ethics, computer ethics, and environmental ethics have flourished with this model of technological development. The problem is not that the framework prevents the lens of ethics from being brought to bear. Rather, the problem is that the framework pushes the processes by which technology is developed out of sight; it turns attention away from technology while it is still in-the-making – while its meaning and material design are in the process of being set.

---

[1] This is the framework I presumed in most of my early work on computer and engineering ethics; see, for example, *Computer Ethics* 1st edition, Prentice Hall, 1985.

In short, the framework turns the lens of ethics away from the earliest and, arguably, most powerful stages of technological development.

Scholars in the field of science and technology studies (STS) have provided a critique of the social impact framework and introduced alternative models of the processes by which technologies are designed, adopted, modified, and used. The literature emphasizes that technology develops in a social context, by means of social processes, and that technology is not just material objects, but rather sociotechnical ensembles – combinations of artifacts, social practices, institutional arrangements, systems of knowledge, and nature. To say that technologies are sociotechnical is to say that technological endeavors are achieved by combinations of artifacts and social practices. For example, software does nothing on its own; software functions in combination with hardware (e.g. computers, electrical systems, routers) and humans organized in various ways (in organizations, agencies, families, etc.) and behaving in particular ways (e.g. using keyboards, responding to signs on screens, interpreting output).

Since the processes by which technologies are designed and developed and come to have meaning are *social* processes, ethicists and ethical notions can be and often are part of the processes. That is, since technological development is not entirely dictated by nature and since what is developed is a function of social factors and arrangements as well as nature, ethical notions and ethicists can influence what is developed.

This has both normative and descriptive implications for anticipatory ethics. Since technological development is a social endeavor, the endeavor can be intentionally structured so that ethical concerns are taken into account early on. This is precisely the normative mission of the 21st Century Act; it structures the research environment for nanotechnology to ensure that ethical issues are addressed early on. But there are also implications for understanding how technological development occurs and always has happened. That is, social forces, stakeholder interests, politics, and history have always influenced the development of technologies and so have moral notions and practices. Think here of the debate about stem cell research; it is a debate about whether or not and how a moral belief should shape the processes of scientific and technological advancement. And think of the regulations with regard to the use of human and animal subjects in research and how these moral concerns have affected the nature of research.

So, ethical notions and practices and ethicists *do* in fact influence technological developments and *could and should* have a more intentional role in shaping technologies in the future. We can examine the influence of moral notions and practices on past and current technologies and we can consider how best to structure (or restructure) design and development processes so as to give ethicists and moral notions a role in development processes.

With the new understanding of technological development, anticipatory ethics is not just plausible but is an activity that has been ongoing, albeit often below the surface of recognition and somewhat out of the grasp of intentional efforts. Acknowledging that ethicists and ethical notions and practices can and do have a role in shaping technological development is, in some sense, the easy part. The hard part is to figure out how to bring ethical notions and practices and ethicists explicitly, intentionally, and effectively into the fray. The difficulty of the task will now be illustrated by taking up the case of software agents.

5.3 Anticipating Software Agents: An Argument for Moral Ontology

Where might we look to observe moral notions and practices being negotiated in software agent technology? Where might we look for opportunities to normatively influence what is being developed? In the case of software agents, opportunities are not hard to find because at least some of the discourse around this technology is explicitly directed at its moral features and moral implications. This literature clusters around the question whether, or in what sense, software agents (and embodied machines containing software, e.g., robots) can be said to be moral agents (Floridi and Sanders 2004; Allen et al. 2005; Allen et al. 2000; Moor 2006). The issue arises in part at least because autonomy is a central component in traditional notions of moral agency. Philosophically, morality only makes sense for autonomous beings; moral agency is only possible in entities with moral autonomy. Thus, if software agents or robots have autonomy, they are candidates for *moral* autonomy and moral agency. This has led a number of philosophers to entertain the possibility of *artificial moral agents* (or AMAs) (Floridi and Sanders 2004; Moor 2006; Wallach and Allen 2009).

The theoretical possibility of artificial moral agents is embedded in a discourse of artificial intelligence, computational modeling, and cognitive science. If computation can "unlock the mysteries of the universe" by modeling reality, then it ought to be able to model morality. In other words, if morality is comprehended by humans through their intelligence and cognition, then *artificial* intelligence and computational cognition should be able to model morality. The model can, then, be embedded in software and hardware to produce entities that behave in ways that are comparable to human moral behavior. These entities, according to the argument, will be artificial moral agents.

The interest of computational modelers in developing artificial moral agents has converged with a more pragmatic, computational endeavor to program machines to behave morally. The pragmatic endeavor is to ensure that when software makes decisions, the decisions it makes (and the consequent behavior) accord with morality. *Machine Ethics* is the term being used for this activity. Anderson and Anderson (2006) describe the goal as follows:

> A goal of machine ethics is to create a machine that's guided by an acceptable ethical principle or set of principles in the decisions it makes about possible courses of action it could take. The behavior of more fully autonomous machines, guided by such an ethical dimension, is likely to be more acceptable in real-world environments than that of machines without such a dimension. (Anderson and Anderson 2006, p. 10)

So two different interests – one in modeling morality as an exercise in artificial intelligence and computational cognition and the other in building decision making devices that incorporate moral principles – converge on a set of questions about the status, meaning, and significance of software agents. The convergence has generated a rich discourse, though it is wide-ranging both because it draws from a variety of disciplines and because it spans a spectrum from highly speculative futuristic visions to concrete programming strategies.

The discourse has not eschewed discussion of responsibility and accountability because in moral theory, notions of agency and autonomy are intertwined with moral responsibility. Individuals have moral responsibility for their behavior in virtue of their having autonomy. If individual behavior were entirely controlled by factors outside the individual or outside the control of the individual, then the individual could not be held morally responsible for their behavior. In this context the autonomy of software agents is crucial to claims about their moral agency. As well, in this context, the move to locate moral responsibility in the software agent seems plausible if not necessary.

The convergence of interests and the resulting discourse around software agents takes us back to the starting place of this chapter, the potential for a collision between the development of software agents and notions and practices of accountability. The discourse around software agents is a discourse about the meaning, significance, and design of a technology. It is about what to "make" of software agent technology – how it should be understood, what features it should be understood to have, and what role it should have in human lives. The collision course concern is that the construction of software agents as autonomous (moral) agents "makes" them something that humans will not be able to understand and control. That they are autonomous may mean that they will be – to some extent at least – beyond human comprehension and control. It is this construction of software agents that seems likely to collide with notions and practices of accountability.

The question is, then, whether anticipatory ethics can or should intervene to avoid a collision or to fit moral notions and practices to what is being developed. As will be illustrated in a moment, it is in the nature of anticipatory ethics that we can't be sure that the collision will take place without intervention; hence, the question whether anticipatory ethics should intervene is by no means simple.

In the remainder of this chapter I will put forward and defend an argument of the kind that seems to be called for by anticipatory ethics. It is an argument for rejecting the characterization of software agents as autonomous. The argument was initially introduced in Johnson and Miller (2008). Here it is extended and elaborated as a way of exploring the promises and pitfalls of anticipatory ethics. To be sure, it is an odd argument since individuals can use words as they wish and they would not be using "autonomous" if its use did not achieve some useful purpose. Although odd, the argument is, nevertheless, important because it does precisely what is sought in anticipatory ethics. It aims to influence the technological endeavor early on by influencing the construction of the meaning of software agents. One of the most powerful ways to change a process and its outcomes is to change the understanding of the endeavor. A change in the understanding of what is being sought ultimately changes what is produced. Thus, a change in the conceptualization, understanding, and discourse of software agent technology would be a significant outcome of anticipatory ethics.

5.3.1 The Argument

In essence, the argument is an argument for a moral ontology.[2] Since the core idea of anticipatory ethics is that moral concerns be taken into account early on in a technology's development, the argument is that accountability issues should be used in conceptualizing what the technology is and identifying its distinctive significance.[3] Software agents ought to be understood – ontologically – as human-made components of sociotechnical systems. They ought to be understood as components of systems constituted and deployed by humans, for human purposes. Software agents function as a result of combinations of humans and artifacts working together. Constructing software agents as autonomous gives the technology an ontological status that disconnects its behavior from those who design and put it into use. Conceptually tethering software to the humans who design and use it constrains the temptation to move the locus of accountability for the harmful effects of software agent behavior away from humans to software agents (Johnson and Miller 2008).[4]

5.3.2 Anticipating Accountability

The argument gains support from an analysis of notions and practices of accountability. At the most basic level, systems and practices of accountability involve the idea that individuals and collectivities (organizations, companies, agencies, countries) are expected to behave in particular ways – according to norms, standards, expectations, or principles. When an individual or a collectivity fails to adhere to a norm, the individual or organization is expected to explain, that is, to give an account, and, depending on the account given, the individual or organization may be liable to certain consequences – shame, mistrust, punishment, compensation, further scrutiny, etc.

Accountability works both retrospectively and prospectively. In retrospective accountability one is held to account for one's actions (or inactions) after they have

[2] I am grateful to Martin Anderson for first characterizing the argument in this way (as moral ontology) while it was still somewhat inchoate in my thinking.

[3] The argument is inspired by Bowker and Starr (1999) and other work that points to the powerful effects of systems of classification.

[4] Of course, software behavior causally contributes to events with untoward consequences. See Johnson and Powers (2005). The locus of accountability is connected to but different from causality.

occurred. We most often see retrospective accountability operating when something untoward has happened, something, that is, that was not supposed to happen. Thus, in the aftermath of Hurricane Katrina, FEMA was held accountable for not responding quickly to the event. When Bernie Madoff's ponzi scheme was uncovered, Madoff was held retrospectively accountable for years of deceitful, exploitative, criminal behavior. In both cases consequences followed, though the nature of the consequences varied. FEMA's Director was fired; public trust in the organization was diminished. Madoff's retrospective accountability involved arrest, trial, testimony, and jail.

Prospective accountability is future looking; it involves practices that inform and remind individual or institutional actors that they are expected to behave in certain ways. In certain domains of life, prospective accountability has an added dimension; individuals and organizations are formally required to demonstrate, by giving an account of some kind, that they are adhering to rules or fulfilling their responsibilities or doing what should be done to prevent untoward events from happening in the future. The most salient examples of this aspect of prospective accountability are in institutional accountability. Employees are asked to fill out conflict of interest statements to demonstrate that they are not in relationships that might bias their decisions. Public companies are required to provide reports to demonstrate that they are fulfilling responsibilities to stockholders.

These two forms of accountability work together. Prospective accountability is aimed at preventing the occurrence of incidents or events that would call for retrospective accountability. Likewise, retrospective accountability supports prospective accountability in the sense that when individuals and organizations are retrospectively held to account, it demonstrates to others that they are accountable for their behavior; that is, it reminds others that if they don't behave according to norms or expectations, they may have to account, retrospectively, for their behavior.

The argument for a moral ontology for software agents is an argument for conceptualizing the technology in a way that will facilitate the operation of prospective and retrospective accountability. The argument is especially compelling because of the current state of legal accountability (liability) for software. In the US at least, legal liability for harmful effects resulting from the use of software is highly uncertain. At best it is a patchwork of generic laws and extremely varied case law drawing on contracts, strict liability, negligence, misrepresentation, limited warranties, and unconscionable disclaimers (Zollers et al. 2005; Ballman 1996; Terry 2002; Childers 2008). Generic laws applying to products and services apply to software but software defies attempts to fit it to one of these categories and prevailing law or legal precedents depend on this distinction. There is no major legislation in this domain and no major legal decisions have broadly addressed software liability. In 2009 the American Law Institute (ALI) approved the final draft of Principles of the Law of Software Contracts but the document testifies to the complexities of liability within software contracts. Thus, currently software developers have little to go on to anticipate their liability in the event that their software causes untoward consequences. Other than the broadest principles of legal liability, there is nothing certain about retrospective or prospective accountability. Yet computer scientists are well aware of the risks in software and they are pervasive and powerful.[5]

5.4 Anticipating Software Agents: The Counterarguments

5.4.1 The Concern Is Premature

An important counter to the argument for a moral ontology is the claim that it is too early to interfere with the development of software agent technology. It is premature – some might say – to stop the use of what is a very useful metaphor, and, anyway, sooner or later – so the argument goes – issues of accountability will be addressed.

This is an important counter because it is, at least in certain respects, consistent with the descriptive implications of the model of technological development described above. The development of software agent technology, like all technological development, is not just socially embedded, the trajectory of development is fluid and contingent. This means that any number of factors may come into play at any time and it means that it is possible and perhaps likely that ethical concerns and issues of accountability will eventually arise and be addressed. At some point or another, the public, the law, or politics will respond to what is being developed. If software agents are deployed in situations in which they put individuals at undue risk or have harmful effects, there will be a response and practices of accountability will be worked out.

Although Wallach and Allen (2009) do not explicitly make this argument in *Moral Machines*, they implicitly adopt the strategy; they predict that in the short term "product safety laws will continue to be stretched to deal with artificial agents" and that dangerous practices will be dealt with first by the courts and later by legislation. They predict that "companies producing and utilizing intelligent machines will stress the difficulties in determining liability and encourage no-fault insurance policies" (p. 198). So Wallach and Allen seem to think that issues of accountability will arise and be addressed sooner or later. Technological development is a social endeavor embedded in society and so social and ethical norms are likely already in play or will come into play at some point or other.

However, the normative thrust of anticipatory ethics is that it is better to take ethical concerns into account *sooner rather than later*. The argument from moral ontology does precisely that; it proposes that while software agent technology is still "in the making," the endeavor to create it should not be understood as an endeavor to create discrete and autonomous entities, but rather to create sociotechnical systems,

[5]Discussion of the risks from defects and failure of software can be found in The Risks Digest, a Forum On Risks To The Public In Computers And Related Systems; the Forum is an activity of the *ACM* Committee on Computers and Public Policy, moderated by *Peter G. Neumann* and found at: http://catless.ncl.ac.uk/Risks/.

i.e., systems that work through the combination of human and non-human components. Such a re-conceptualization does not prevent explanations of software and hardware behavior, it prevents the complete deflection of human responsibility for harmful effects resulting from software behavior.

So, is it better to intervene in the construction of software agents early on or better to let the development process unfold without constraining the way the technology is conceptualized and understood? This is not a simple question. Addressing ethical issues early on has advantages; it also has disadvantages. Bringing in moral concerns early on may blunt efforts that would have been fruitful and might have turned out to be morally unproblematic. On the other hand, not bringing in these concerns early on might lead to technologies that run into trouble with the public, the law, or other stakeholders later on. The development process is *not* – as it is understood in technological determinism – a process in which a particular or pre-determined entity is waiting to "emerge." In the technologically deterministic view, all we need to do is make sure that the environment for development is unencumbered so that the pre-determined entity can emerge quickly. On the contrary, bringing ethical issues into technological development has promise precisely because the development process is fluid and contingent.

Research and development take place in particular places, by particular individuals and groups and the when, where, how, and who is involved make a difference, just as the amount of funding and a myriad of other factors make a difference. Anticipatory ethics and the argument for a moral ontology should not, then, be seen as an encumbrance to development. Taking issues of accountability into account early on will make a difference and there may even be trade-offs in doing so, but so it is with all the factors that influence technological development.

The parallel between software agent technology and ethical issues in other domains of research is helpful here. Consider the moral constraint that has been institutionalized for medical research involving human subjects. When the requirement that scientists doing medical research obtain the informed consent of human subjects was first instituted, it was felt, by some scientists at least, to be a constraint on their research. Some thought that the science would proceed more quickly and more effectively if scientists didn't have to obtain the informed consent of human subjects. The requirement is now well accepted in medical research and it seems fair to say that what is discovered and developed in medical research involving human subjects is different than what might have been learned otherwise. Some would say the science is better for it; others might not agree. And, of course, it depends on what you mean by "good" science.

The lesson, it would seem, for anticipatory ethics and the development of software agent technology is that conceptualizing the technology in ways that keep it tethered to the humans who design and use it will make a difference.

5.4.2 Software Agents Are Autonomous

But what about the counterclaim that software agents *are* autonomous? Isn't it a misconception to think that software agents are not autonomous? They operate independently – in space and time at least – from the humans who design and deploy them! Oddly, no one seems to claim exactly this; that is, no one seems to claim that the terms autonomy and autonomous agent have a singular, objectively definable meaning. These terms are being used in a wide variety of different ways. Some may deny my characterization of their use as metaphorical; others may embrace it. Computer scientists and software developers seem to use the terms as a metaphor that helps them describe how software agents operate in a simple way that does not require technical expertise. In some sense, computer scientists and software developers may not have a serious stake in the use of these terms since they understand what they are doing in a technical disciplinary discourse, be it programming, software development, or electrical engineering.

Whatever their interests, software developers have quite different stakes in the metaphor than do philosophers and cognitive scientists. In fact it would seem that philosophers and cognitive scientists use autonomy and autonomous in technical ways, technical in the sense that they are embedded in philosophical theory. One of the most influential pieces on the topic of artificial agents argues for autonomy as an operationally defined term tied to a particular level of abstraction (Floridi and Sanders 2004). Others seem to make equivalency claims, that is, they claim that machines will be moral agents in the sense that they will have features that are equivalent to those of human moral agents. Of course, it is far from clear what constitutes equivalency in this context, and it is not at all clear what the significance is of an "entity" that has autonomy at a specific level of abstraction.

These different uses of "autonomy" and "autonomous" contribute to the creation of a rich discourse, a discourse with the potential for creative thinking and fruitful cross fertilization among disciplines and theoretical frameworks. On the one hand, the discourse often seems confused and misleading as terms are moved from one context to the next, are used in widely different ways, and interlocutors often seem to miss one another. Do the issues need to be so complex? The discourse is complex in part because it involves an ontological struggle. The discourse is about what we are to "make" of what is being developed.

Suppose we try to make it simpler. Why not understand "autonomous agents" simply to refer to things that operate on their own. It is easy enough to think about refrigerators, automatic pilot systems, and search engines (all of which today are in part constituted with software) as operating on their own. When pressed, however, it is not so easy. That is, when we try to justify this easy account, it becomes clear that we are engaged in ontology. To consider refrigerators, automatic pilots, search engines or software as entities – not even agents, just entities – is to engage in the mental exercise of separating them out from the complex sociotechnical systems of which they are a part.

Yes, my refrigerator maintains its internal temperature "on its own"; the thermostat signals other components to behave in certain ways that raise and lower the temperature. The problem with saying this is that my refrigerator only works insofar as it is plugged into an enormously complex power grid, a power grid that depends on many human and non-human components. In fact, the institutional arrangements constituting the power grid are an enormous feat of human social cooperation and interdependence. And, of course, my refrigerator only works as it is supposed to if I buy food that needs to be refrigerated, open and close the door to put the food in and take it out, pay my utility bill, and so on. Where does the entity that I call "my refrigerator" begin and end? It seems that we have decided (perhaps arbitrarily, perhaps not) to draw lines. We have conceptually (abstractly) decided we will count the rectangular chunk of plastic and metal that sits in my kitchen as "a refrigerator." We have decided to leave on the other side of the line (outside of the concept) the electrical grid to which it must be connected, all the people who maintain the electrical grid, and my behavior in opening and closing the door to put in and take out food. Yet my "refrigerator" does not work as a refrigerator unless all these other human and non-human components do their part. My refrigerator is a sociotechnical system, and the hunk of metal and plastic that I brought home from an appliance store is merely one component of that system.

It's the same for the automatic pilot. The automatic pilot works only insofar as it is delicately connected to other parts of the airplane. Whether the automatic pilot goes on only when human pilots flip a switch or goes on under specific conditions without human action is a design feature chosen by humans. What the automatic pilot does is the result of interactions among human and non-human parts. Where, again, does the automatic pilot begin and end? Is it an entity in itself or a component of an airplane? The airplane itself is a component in an enormously complex air transportation system. We draw lines; the lines specify what an automatic pilot "is"; what an airplane is, and so on. We choose, that is, what we will conceptualize as the part and what we will conceptualize as the whole.

It is the same for software. We say that a set of lines of code is software. Some call the software an agent. Of course, the software does nothing unless it is put into machines. Some call the software and hardware together an agent. Of course, humans had to create the software and humans had to build the machines and embed the software in the hardware. Humans turn on the machines, test and monitor their behavior. Where does the software agent (or the robot) begin and end? Some may argue that there is something different and distinctive about computers and software – they are not just chunks of metal and plastic; they are computational. This, they will say, makes them closer to or the same as humans. But this is another line drawing matter.

Lines do, of course, have to be drawn and they are drawn for various purposes. The thrust of the argument here is not to deny this, but to argue for bringing moral considerations into our line drawing. The argument for a moral ontology is an argument to draw the lines of "software agents" with an eye to keeping the locus on accountability with humans.

In this context two different sorts of dangers seem at issue. One has already been identified, that software entities might be conceptualized so as to suggest that no humans are accountable for the behavior of the software agents. The other is that theory- or context-dependent notions will move from one context to another in ways that cause confusion and are misleading. Grodzinsky et al. (2008) illustrate this when they use the Floridi and Sanders (2004) notion of autonomy at different levels of abstraction. Focusing on tables used in programming, they show how software can be autonomous to users (who cannot modify the table) while at the same time not autonomous to the designers (who can modify the table).

Thus, the argument for moral autonomy cannot be countered by the claim that software agents just are autonomous. Software agents are sociotechnical systems. They depend for their operation on being embedded in larger, more complicated systems that function by combinations of human and non-human or artifactual activity. Conceptualizing software agents as sociotechnical systems might well make a moral difference.

5.5 Conclusion

There are several lessons to be derived from the preceding analysis. The first is that anticipatory ethics is tied to a view of technological development as a fluid and contingent social endeavor and, hence, one that can be influenced by ethics and ethicists. This view has both descriptive and normative implications; it reveals both that moral notions and practices may have been at work influencing technological development in the past and influencing the development of existing technologies *and* that they can be more intentionally and effectively brought into play in the development of new technologies. In the case of software agent technology, the analysis focused on an argument for a moral ontology for software agents. The argument claims that we ought to conceptualize and understand software agents in ways that keep them tethered to the humans who design and deploy them, so as to avoid a deflection of human responsibility for the behavior of software agents. The claim is that an ontology of this kind will allow prospective and retrospective accountability to work. Although the argument illustrates the promise of anticipatory ethics, in the end it seems that anticipatory activity must be viewed cautiously. There is no certainty, for example, that eliminating the characterization of software agents as autonomous is the only or best way to address issues of accountability. Since technological development is fluid and contingent there are any number of ways that moral norms and practices can come into play. What is clear, nevertheless, is that technologies including software agents are sociotechnical systems and while we can conceptualize them in other ways, doing so can be misleading and may get in the way of human accountability.

References

Allen, C., I. Smit, and W. Wallach. 2005. Artificial morality: Top-down, bottom-up, and hybrid approaches. *Ethics and Information Technology* 7: 149–155.

Allen, C., G. Varaner, and J. Zinser. 2000. Prolegomena to any future artificial moral agent. *Journal of Experimental and Theoretical Artificial Intelligence* 12 (3): 251–261.

Anderson, M., and S.L. Anderson. 2006. Machine ethics? *IEEE Intelligent Systems* 21 (4): 10–11.

Anderson, M., and S.L. Anderson. 2007. The status of machine ethics: A report from the AAAI symposium. *Minds and Machines* 17: 1–10.

Ballman, D.R. 1996. Commentary: Software tort: Evaluating software harm by duty of function and form. *Connecticut Insurance Law Journal* 3: 417.

Bowker, G.C., and S.L. Starr. 1999. *Sorting things out: classification and its consequences*. Cambridge, MA: The MIT Press.

Chien, S., R. Sherwood, D. Tran, B. Cichy, G. Rabideau, and R. Castano. 2005. Lessons learned from autonomous sciencecraft experiment. Paper presented at the autonomous agents and multi-agent systems conference, Utrecht, Netherlands.

Childers, S.J. 2008. Don't stop the music: No strict products liability for embedded software. University of Florida Journal of Law & Public Policy 19: 125, 127.

Fisher, E., C. Selin, and J.M. Wetmore. 2008. *The yearbook of nanotechnology in society vol. 1: Presenting Futures*. New York, NY: Springer.

Floridi, L. and J.W. Sanders. 2004. On the morality of artificial agents. *Minds and Machines* 14 (3): 349–379.

Grodzinsky, F.S., K.W. Miller, and M.J. Wolf. 2008. The ethics of designing artificial agents. *Ethics and Information Technology* 10: 115–121.

Johnson, D. and J.M Wetmore 2008. STS and ethics: Implications for engineering ethics. In *New handbook of science, and technology studies*, eds. E. Hackett, O. Amsterdamska, M. Lynch, and J. Wajcman. Cambridge, MA: The MIT Press.

Johnson, D. and K. Miller. 2008. Un-making artificial moral agents. *Ethics and Information Technology* 10 (2–3): 123–133.

Johnson, D. and T.M. Powers. 2005. Computer systems and responsibility: A normative look at technological complexity. *Ethics and Information Technology* 7 (2): 99–107.

Joy, B. 2000. Why the future doesn't need us. *Wired* 8 (4): 238–262.

Lee, M. H. and N.J Lacey. 2003. *Minds and machines* 13: 367–395.

Moor, J. H. 2006. The nature, importance, and difficulty of machine ethics. *IEEE Intelligent Systems* 21 (4): 18–21.

Noorman, M. 2008. *Mind the gap a critique of human/technology analogies in artificial agent discourse*. Universitaire Pers Maastrict,

Rahwan, I., L. Sonenberg, N. Jennings, and McBurney, P. 2007. Stratum: A methodology for designing heuristic agent negotiation strategies. *Applied Artificial Intelligence* 21 (6): 489–527

Terry, N.P. 2002. When the "machine that goes 'ping" causes harm: default torts rules and technologically-mediated health care injuries. *Saint Louis University Law Journal* 46: 37–59.

The American Law Institute. 2009. Principles of the law of software contracts, proposed final draft 16 Mar.

Wallach, W. and C. Allen. 2009. *Moral machines: teaching robots right from wrong*. Oxford: Oxford University Press.

Zollers, F.E., A. McMullin, S.N. Hurd, and P. Shears. 2005. No more soft landings for software: Liability for defects in an industry that has come of age. *Santa Clara Computer and High Technology Law Journal* 21: 745.

Chapter 6
Sui Generis Rules

Lyria Bennett Moses

There is the story of a Vermont justice of the peace before whom a suit was brought by one farmer against another for breaking a churn. The justice took time to consider, and then said he had looked through the statutes and could find nothing about churns, and gave judgment for the defendant.
– Oliver Wendell Holmes (1897)

6.1 Introduction

Although the "pace" of technological change is difficult to measure (Edgerton 2006), few would deny that technological change is a persistent feature of our society. When new technologies are introduced into society, and begin to spread, there is often pressure on the legal system to "respond" or "keep pace" (Bennett Moses 2007a). As technology changes, new entities, activities and relationships become practical possibilities. As a result, there may be (1) pressure to enact new laws, (2) a need to resolve uncertainties as to the application of law in new contexts, (3) legal rules that apply poorly in new contexts when measured against achieving an underlying goal, and (4) laws that can no longer be justified and hence become obsolete (Bennett Moses 2007a). When contemplating law reform in response to the first three problems, there is, not surprisingly, a tendency to propose new rules designed to apply to the new entities, activities and relationships that gave rise to the difficulty. In other words, there is a tendency to treat new entities, activities and relationships relating to new technologies as in need of special, or *sui generis*, regulation or protection.

As the enactment of *sui generis* rules is one common temptation for rule-makers wishing to respond quickly to emerging technologies, this chapter will consider the extent to which *sui generis* rules are an effective approach for dealing with the pacing problem. In deciding whether *sui generis* rules are truly appropriate,

L. Bennett Moses (✉)
University of New South Wales, Sydney, NSW, Australia
e-mail: lyria@unsw.edu.au

G.E. Marchant et al. (eds.), *The Growing Gap Between Emerging Technologies and Legal-Ethical Oversight*, The International Library of Ethics, Law and Technology 7,
DOI 10.1007/978-94-007-1356-7_6, © Springer Science+Business Media B.V. 2011

it is important to take account of their advantages and disadvantages compared to a more broadly-framed approach. It is also necessary to consider alternative approaches, such as utilizing a broad category and tailoring the law's application to the new entity, activity or relationship within that broad category or employing relatively technology-neutral *sui generis* rules. Unless the tendency to enact narrowly framed technology-specific legislation is minimized, the possibility for further legal problems as technology continues to evolve is high.

6.2 *Sui Generis* Rules: Special Laws for Special Circumstances

As I have discussed elsewhere (2007b), technological change creates significant problems for law. Existing laws are often subject to uncertainty in their application to new situations or, if certain, may not apply as intended. In addition, it is often felt that new laws are necessary, for example to regulate a new technology. As a result, it is often the case that entities, activities and relationships made possible through technological change come to be governed by *sui generis* rules. The term "*sui generis*" means "of its own kind" (Oxford English Dictionary). Thus laws are *sui generis* to the extent that they treat a particular entity, activity or relationship as subject to a narrowly crafted legal regime.

The fact that particular entities, activities and relationships are subject to a narrowly tailored legal regime is sometimes inevitable. There are situations where there is no broader category under which the goal sought to be achieved by particular rules could be achieved. In these situations, the fact that *sui generis* rules are employed is not a matter of choice (except to the extent there is a choice to regulate at all). An example of a rule that could only be crafted narrowly is section 14(1) of the *Apiaries Act 1985* (NSW, Australia), which provides "A person shall not keep bees, or allow bees to be kept in an apiary, except in a frame hive." If the legislature wishes to mandate frame hives for beekeeping, a specific law is the best means of doing so. One might ask whether legislative intervention is justified on this issue, but there is little question of drafting the law more broadly. Sometimes a law, initially *sui generis*, will come to be seen as falling within an as yet undiscovered broader category. For example, copyright law originally consisted of narrowly tailored legislation such as *An Act for Encouraging the Art of Making New Models and Casts of Busts, and other Things therein Mentioned*, 38 Geo. III c. 71 (1798) (UK). At the time the Act was passed, there was no broadly applicable category of copyright and thus sculpture was protected through a *sui generis* statute. Later, the more broadly crafted copyright legislation subsumed the *sui generis* protection that had been offered to particular modes of creative expression.

Often there is a choice between adopting a *sui generis* regime to deal with a particular problem and regulating conduct through an existing, broadly framed, law. For example, the advent of genetic testing required a decision as to whether genetic information should be subjected to *sui generis* privacy laws or included within a broader category of protected information. Different jurisdictions reached different conclusions on this question. In Australia, privacy protection is granted in a general, rather than *sui generis*, law. In section 6 of the *Privacy Act 1988* (Cth), "health information" is defined to include "genetic information about an individual in a form that is, or could be, predictive of the health of the individual or a genetic relative of the individual." Further, "sensitive information" is defined to include both "health information" and "genetic information about an individual that is not otherwise health information." Thus the privacy of genetic information is protected by incorporating it in one of two more general protected categories of information. The opposite approach is taken in some other jurisdictions, such as Delaware. Chapter 12.II of Title 16 of the Delaware Code contains specific provisions regulating the taking and disclosure of genetic information.[1] In the case of genetic testing, a new technology generated a new type of information (genetic information) that many felt needed to be protected by privacy laws. Each jurisdiction had a choice as to whether to protect genetic information through the enactment of a *sui generis* law, or by including genetic information within a broader class of protected information, thus making it subject to a broadly crafted privacy regime.

Although modern intellectual property law consists primarily of broadly crafted categories (patents, copyright, trade marks), there remain pockets of *sui generis* protection. The Semiconductor Chip Protection Act was enacted in the United States 1984. It was designed to protect the semiconductor chip industry from reverse engineered copies. The Act was narrowly crafted to protect "mask works," being the design element in semiconductor chips. At the time, the Act was widely lauded as an effective response to an industry's need for intellectual property protection in light of the under-inclusiveness of existing regimes (e.g. Samuels and Samuels 1986; Michaelson 1986). Special protection for semiconductor chips is now mandatory for all members of the World Trade Organisation through Article 35 of the Agreement on Trade-Related Aspects of Intellectual Property Rights (TRIPS), and many countries have thus enacted *sui generis* legislation.

Sometimes, an entity, activity or relationship is treated *sui generis* in the absence of any specific legislative enactment. Human in vitro embryos are an example of this. Such embryos were first created in 1969 (Edwards et al. 1969). In theory, they could fall under the law of persons or the law of property. However, in *Davis v. Davis*, it was held that "preembryos are not, strictly speaking, either "persons" or "property," but occupy an interim category that entitles them to special respect because of their potential for human life." It is not clear from this decision whether this implies (1) that embryos are special property, able to be treated as objects of property rights but subject to constraints necessary to ensure respectful treatment, or (2) that embryos cannot be objects of property rights. If embryos are neither persons nor property, interactions with embryos are regulated, if at all, through *sui generis* rules.

---

[1]The Delaware law remained applicable following passage of the Genetic Information Non-Discrimination Act of 2008 (see section 209 of that Act).

6.3 *Sui Generis* Rules and Other Dichotomies

Laws may be measured on a range of scales. As discussed in the previous section, laws may tend to be tailored to a narrow range of circumstances (*sui generis*) or intended to apply to broader categories of entities, activities or relationships. There are various other criteria by which laws might be compared, including the distinction between rules and standards, between laws that discriminate between technologies and those that do not, and between legislation that is technology-neutral and legislation that is not. Each of these overlaps with the distinction drawn here between *sui generis* and broadly-framed laws, but none is identical.

There is an old and often used distinction between laws that are rule-like and laws that are standard-like. Laws are considered rule-like to the extent that they are given content before individuals act (Kaplow 1992). Because the distinction between rules and standards is usually based on the content of a law rather than its scope, both rules and standards can be *sui generis* or broadly applicable (Schlag 1985). The distinction between *sui generis* and broadly crafted laws thus does not map onto the distinction between rules and standards, as traditionally conceived.

Although laws that are *sui generis* often overlap with those that discriminate between different technologies, the two categories are not identical. Laws can be said to discriminate between technologies where they treat different technologies differently even where the technologies produce equivalent results (e.g. van der Haar 2007). In so far as *sui generis* rules create special laws for particular circumstances, a side effect may be differential treatment of similar situations, possibly based on differences in the technology used. However, even laws electing to employ general categories can discriminate between technologies. The Australian Privacy Act, discussed above, does not treat genetic information as *sui generis* but rather includes it within broader categories of protected information ("health information" and "sensitive information"). However, information obtained from proteins in the blood, rather than through genetic analysis, will not necessarily be "health information" or "sensitive information." Similar information is thus treated differently depending on the technology used to obtain it (genetic testing or proteomics). This is true despite the fact that genetic information is not treated *sui generis*. Thus while *sui generis* rules may lead to discrimination between equivalent technologies, the narrowness of rules does not map directly to their discriminatory effect.

One dichotomy that comes up in the context of legislating in contexts of rapid technological change is the distinction between technology-neutral and technology-specific rules. This terminology has a range of potential meanings (Koops 2006; Reed 2007), but here the term "technology-neutral" is used to signify laws that are designed to be independent of any particular technological context so that they can continue to apply appropriately even as technology changes. Again, there are overlaps between my distinction between *sui generis* and broadly crafted rules and the distinction between technology-specific and technology-neutral rules. The Semiconductor Chip Protection Act is an example of legislation that is both *sui generis* and technology-specific. But it is possible to envisage laws that would fall into one category but not the other. The special status of human in vitro embryos might be independent of the technology of their creation and storage. On the other side, the Australian Privacy Act does not treat genetic information as *sui generis*, yet it does assume that familially-linked information about an individual's future health risks will be extracted through genetic testing rather than other means. Information obtained through another means, such as proteomics, would not automatically be treated as "health information" under the Privacy Act unless obtained in a context specifically referred to in the legislation.

The issue of whether there is a need for *sui generis* rules is different than the question of the necessity for *sui generis* ethics along the lines of computer ethics. Within applied ethics, there is a debate about the "uniqueness" of a subject-specific field such as computer ethics. There are those who suggest that ethical issues in this area are unique in that they could not have arisen before the advent of the technology (e.g. Maner 1996). In contrast, Johnson (2001) has pointed out that while the specific issues may be unique, their solutions can be found by adopting an established moral framework. Himma (2003) has observed the irrelevance of this debate about uniqueness to the question of whether "computer ethics" deserves to be treated as a specialized field of applied ethics. In law, there is a similar debate about the usefulness of subject-specific analysis, such as "cyberlaw" (compare Lessig 1995, 1999; Easterbrook 1996; Sommer 2000). My focus here is not on the question of whether new technologies raise new legal issues (Bennett Moses 2007a) or whether these legal issues ought to be the focus of specific study, but rather on the question of how such legal issues ought to be resolved. In particular, this chapter asks in what circumstances the appropriate response is the creation of a legal rule designed to apply only in a narrow range of circumstances.

6.4 Why Employ *Sui Generis* Rules?

There are many reasons rule-makers might choose to craft legal rules narrowly. Sometimes, a narrowly tailored rule is the only available means of achieving a particular goal. Yet, in many circumstances, there is a choice between regulating an entity, activity or relationship through a broader regime (such as patent law or property law) and creating a more narrowly tailored law. In such cases, the reason for choosing the latter will usually be a real or perceived difference between the broader and narrower subject matter.

For example, many have argued that the nature of genetic information, such as its sensitivity, connectedness to family and predictive nature, justify *sui generis* treatment, apart from more general laws protecting privacy and confidentiality (Annas et al. 1995). Others have argued against treating genetic information as *sui generis* (Australian Law Reform Commission 2003; Murray 1997; Gostin and Hodge 1999). The debate generally focuses on the extent to which genetic information is truly distinguishable from broader categories of health information and the risks and benefits of special treatment.

The Semiconductor Chip Protection Act was designed to fill a "gap" between copyright and patent law (Samuelson 1985 at 510–511). At the time, it was felt that copyright law did not apply to objects that were useful in themselves, such as semiconductor chips, and the level of ingenuity required for patent protection would generally be absent (McKeough 1986; Samuelson 1985, H.R. Rep. No. 781). In addition, the time taken to obtain a patent, where available, was longer than the average life-cycle of a particular chip design (Fitz Simons 1990; Samuelson 1985). Accordingly, it was the view of the relevant House Committee that a *sui generis* approach would be best adapted to the needs of the semiconductor industry (H.R. Rep. No. 781).

The problems faced by the semiconductor chip industry are not unusual. Generally speaking, different industries have different average costs of research and development, different average development timelines, different manufacturing costs and infrastructure and different piracy risks (Burk and Lemley 2003). Assuming patent law is intended to promote the development and proliferation of beneficial technologies at minimum cost to society in terms of the content and length of any monopoly granted, an ideal patent law for the pharmaceutical industry might look very different than an ideal patent law for the machine tool industry (Burk and Lemley 2003). Having a single patent regime that applies across the board thus imposes a "uniformity cost" (Carroll 2006). Accordingly, it is not surprising that arguments for *sui generis* protection are not confined to the semiconductor chip industry. Arguments have been made for *sui generis* treatment of the intellectual property of various industries including software (Abramson 2002; Phillips 1992; Samuelson 1985; Samuelson et al. 1994; cf. Raskind 1986; Griem 1993), proteomics (Williams 2005) and biotechnology (Burk 1991; Purvis 1987; Ellinson 1988; cf. Mellor 1988). Although the need for industry-specific *sui generis* patent rules is sometimes questioned (Shi 2005), there are obvious advantages in laws that take account of relevant features of a specific industry.

The decision to treat human embryos as *sui generis* rather than as objects of property has been made by both courts and legislatures. In the United States, *Davis v Davis* stands as a much-cited authority for the proposition that human in vitro embryos are not property. In the United Kingdom, the *Human Fertilisation and Embryology* Act places embryos within a narrowly tailored legal regime. Either way, there are many reasons why it might be thought necessary to treat embryos as falling outside the scope of property law (cf. Bennett Moses 2008). For example, ordinary chattels can generally be sold, yet most people would feel uncomfortable if similar trading in human embryos were permitted. If classifying embryos as part of the law of property would lead to unrestricted trade, it seems better to treat embryos *sui generis*.

In a sense, each of these examples is part of a broader phenomenon. Any law that operates broadly will apply imperfectly in at least some contexts. There is usually some variety in contexts that makes a law seem inappropriate or insufficient some of the time. Even where there is a common goal, such as avoiding unwanted advertising, there are differences in media that justify different treatment for faxes and email (compare 47 U.S.C. 227 and 15 U.S.C. 7701-7713). Similarly, the risks and costs of a single activity such as providing law enforcement authorities with access to telecommunications traffic data might be different in different contexts (Escudero-Pascual and Hosein 2004).

6.5 Dangers of *Sui Generis* Rules

Creating narrowly tailored legal rules has several potential disadvantages. These include the possible failure of those rules to cover sufficient ground, the administrative costs of enacting and maintaining multiple legal regimes, the tendency for *sui generis* rules to assume a temporary technological framework and the potential for narrowly defined legislation to favour narrowly defined groups at the expense of others. The extent to which each of these problems will arise in a particular context will vary – some may be avoided entirely. However, in deciding whether a *sui generis* approach is best, it is important to beware of the potential dangers.

6.5.1 The Problem of Completeness

The first potential problem, the possibility that *sui generis* rules will fail to cover sufficient ground, is the result of a decision to exclude an entity, activity or relationship from a generally operating legal regime. It arises where there is a choice between a generally operating legal regime and better tailored *sui generis* rules, and a decision is made that the latter will replace rather than supplement the former. If the *sui generis* rules fail to cover the same ground as the general regime, there is a risk of gaps and uncertainties.

This problem is evident in the legal treatment of human in vitro embryos. As mentioned above, it was held in *Davis v Davis* that human in vitro embryos were not "property." While the case law is still unclear on the implications of this, the impact of the failure to treat human in vitro embryos as a potential object of property rights is evident in the aftermath of an incident in California. Three doctors associated with the University of California at Irvine were accused of using human embryos in fertilization procedures and research *without the consent of the genetic contributors* (Weber and Marquis 1995). Orange County prosecutors believed that the three could not be charged with "theft" due to the fact that embryos were not "property" (Weber and Marquis 1995; California Penal Code § 503). Ultimately, the only accused doctor remaining in the United States was convicted of federal mail fraud in relation to errors on insurance billing forms (McDonald and Christensen 1998).

Property law is the general mechanism by which the law regulates the interactions between people and "things" (Bennett Moses 2008). The possibility that more than one person might interact with a thing creates a potential for conflict. From a practical perspective, it is the law of property that identifies who is subject to which legal relations with respect to a thing at any given moment and how these can be enforced (Kohler 2000 at 282; Hohfeld 1913–1914). Thus the legal system contains crimes such as theft and torts such as conversion that ensure that objects of property remain under the control of particular people. The legal system does not have rules that perform the same function for things that are not objects of property rights. While *sui generis* rules could perform a similar function in particular circumstances, there is a risk of incompleteness where rules are less comprehensive than property law or there is a delay in their creation (Kohler and Palmer 1998 at 17).

In the case of in vitro embryos, the insistence that embryos are not property has not been matched by a comprehensive *sui generis* legal regime. In the aftermath of the Irvine scandal, the Californian Penal Code was amended to add a more appropriate offence in section 367 g should similar conduct be repeated. However, the provision only addressed the conduct at issue in the scandal (taking embryos and gametes without consent) and only came into effect afterwards. It did not deal with other issues, such as the unauthorised destruction of embryos or negligent handling of embryos. If embryos are not property, tort actions will generally need to be based on emotional distress rather than the law of conversion (e.g. *Del Zio v The Presbyterian Hospital in New York*).

On the other hand, where human embryos are treated as property, remedies are available to protect rights of control. In *York v Jones*, a couple wishing to move embryos from one fertility clinic to another was able to rely on the tort of detinue to recover their embryos. In *Frisina v Women and Infants Hospital*, a claim alleging emotional distress following the loss of embryos was allowed to proceed to the extent that it was based on loss or destruction of irreplaceable property. In *Jeter v Mayo Clinic Arizona*, embryos were recognised as "things," and thus litigation based on breach of bailment and breach of an undertaking to protect "things" was allowed to proceed. In *Dahl v. Angle*, the Court of Appeals of Oregon treated rights of control over embryos as "personal property," which was necessary if the court were to have jurisdiction to make an order respecting the embryos on dissolution of marriage. Property law and related principles thus seem capable of resolving disputes between those with rights to control embryos and those who misuse or damage them. Of course there are disputes about embryos for which property law offers little assistance, such as disputes between different people with rights of control, such as divorcing spouses (Bennett Moses 2005 at 608–615). Nevertheless, property law can provide useful answers in a wide variety of situations.

The example of in vitro embryos illustrates the dangers of removing an entity, activity or relationship from a general domain such as property law without simultaneously creating a *sui generis* regime of similar scope. The problem can be avoided by allowing *sui generis* rules to run in parallel with more generally framed rules. For example, the Semiconductor Chip Protection Act does not prevent the semiconductor industry from accessing patent and copyright protection where each is applicable.[2] The result is that process patent protection continues to apply and has gained in importance as process innovation has increased in importance against chip design (Lewis 1995).

6.5.2 The Problem of Administrative Costs

While a broad and comprehensive *sui generis* regime may be desirable for the reasons outlined above, it can also prove costly. Most obviously, there are costs associated with the creation of legal rules. If created by a legislature, costs are incurred in drafting the new law, analysing its costs and benefits and ultimately enacting it. The main cost is one of time – time spent on subject-specific laws could arguably be better spent on other issues in the jurisdiction. If created by courts, the costs are borne by parties to litigation as well as the court system in hearing and deciding cases where the law is unclear. If there is an alternative existing legal regime that could perform a function similar to the new *sui generis* regime, these additional costs are potentially redundant (Brownsword 2008 at 152).

Depending on the operation of the *sui generis* law, it may envisage the creation of a new bureaucracy to enact accompanying regulations, monitor compliance and enforce the new law. These costs can obviously be reduced if these tasks are delegated to an existing bureaucracy, although the additional tasks may nevertheless result in an increase in size of the bureaucracy.

Another cost associated with narrowly tailored legal regimes is the need for familiarity with different rules for similar but different contexts. For example, if different patent laws apply to different industries (as suggested by Burk and Lemley 2003), intellectual property attorneys need to be familiar with all of them. Companies responsible for inventions in different fields would need to work with multiple legal regimes. Further, the rules designed to determine the regime into which a particular invention fell would inevitably create an entire new field of inquiry with which many would need to be familiar. The result is an inevitable increase in the legal costs of those engaging in research and development.

Not only will there be a need for familiarity with multiple legal regimes, but there will also be more questions of interpretation for courts to decide (Samuelson 1985 at 501–502; Burk and Lemley 2003). Returning to the patent example, if each industry's patent law were drafted using different, tailored terminology, each would have its own questions of interpretation for courts to resolve. In fact, even though different industries share common patent legislation, it has been suggested that the resolution of a patent law question in the context of biotechnology will not necessarily assist in a case involving nanotechnology (Burk and Lemley 2003). If technologies converge, the appropriate legal regime may be difficult to identify. The result may be a rise in the volume of litigation.

[2] 17 U.S.C. § 912(a). See also *J.E.M. Ag Supply, Inc. v Pioneer Hi-Bred Int'l, Inc.* 534 U.S. 124 (2001) (in relation to *sui generis* protection for plants).

Of course, the extent of administrative costs associated with multiple *sui generis* legal regimes depends on how those regimes are crafted. It would be possible to employ similar language across the board in order to reduce the costs associated with drafting and interpreting multiple statutes. It is also possible to employ the same bureaucracy across the field. However, the risk of heavy administrative costs should be taken into account in the decision to create multiple *sui generis* legal regimes rather than a single general framework. Administrative costs are most evident where multiple *sui generis* regimes proliferate within the same regulatory space (see Wahlgren 2004).

6.5.3 The Problem of Technological Change

As demonstrated above, the scales of generality and technology-neutrality are not identical. It is possible to have *sui generis* technology-neutral laws and generally operating, but technology-specific, laws. Despite this, it is common for *sui generis* laws to assume a particular state of technology. In fact, one reason for *sui generis* laws is the need for special laws to deal with an entity, activity or relationship that is the result of technological change (Bennett Moses 2007a).

The Semiconductor Chip Protection Act was enacted specifically to protect an industry built around a new technology from "chip pirates." The Act, which creates a *sui generis* intellectual property regime for semiconductor chips, is highly technology-specific. The Act quickly dated. There has been little infringement litigation employing the Act (Risberg 1990; Callaway 2008). As one chip designer, Hans Camernzind, has noted "Everyone was hoping [the Semiconductor Chip Protection Act] would stop direct optical copying. It doesn't work – nobody's using it, period" (Callaway 2008). The Australian version of the Semiconductor Chip Protection Act has been criticised by the Intellectual Property and Competition Review Committee (2000) for being "highly specialised, technology specific and narrowly defined" and hence unable to keep up with technological change. The reasons for the practical irrelevance of the Act are various but include changing techniques of design and manufacture (Rauch 1993 at 429; Risberg 1990 at 277; Radomsky 2000; Kukkonen 1997 at 133).

To the extent *sui generis* rules assume a particular state of technology, they risk falling behind the times (Bennett Moses 2007a; Brenner 2007; Kirby 2008). Like the Semiconductor Chip Protection Act, they may become obsolete in that the legislation may not apply well to a future manifestation of the technology to which it is directed (see, for example, Breyer et al. 2009, in relation to European regulation of genetically modified organisms). If *sui generis* rules are created to provide differential treatment for a technology considered special in a particular context, it is possible that the technology will change so that it is no longer special or other technologies will develop that are equally special. Legislation drafted in a particular technological mould will not be optimal in either context. Further, difficulties of interpretation and inconsistencies may arise if, as often happens, technologies subject to different *sui generis* regimes ultimately converge (Abelson et al. 2008 at 291; Svantesson 2007 at 45).

Rules that assume a particular technological framework are not only potentially distorting from a legal perspective, they may distort technology as well. Potential avenues for technological change may remain unexplored in order to remain within the technological paradigm assumed by a beneficial law. Alternatively, technology may be redesigned in socially and economically unproductive ways in order to avoid the application of onerous regulation. One technology may be preferred over another equivalent technology due purely to the existence of separate legal regimes for each. Sometimes, encouraging a particular pathway for a technology is the very purpose of regulation, but it can also be an unintended side effect of technology-specific *sui generis* rules.

6.5.4 The Problem of Politics

The final potential problem with narrowly crafted legal rules, at least in the legislative context, is the potential for bias in their negotiation. Although this simplifies things somewhat, broadly applicable rules tend to have more widely distributed costs and benefits than narrowly crafted rules. This makes *sui generis* rulemaking more susceptible to interest group politics and rent-seeking (Wilson 1980). The problem is most easily observed where legislation is designed to apply to a particular industry where powerful players may urge rules that benefit the industry at the expense of possibly as-yet-unknown others. It is evident in the case of *sui generis* intellectual property regimes (Reichman 1994). This is a particular problem in technology based industries where participants may have a monopoly on the information that regulators are using to make decisions (Nelkin 1984). Conversely, public opinion may be galvanised around a specific, narrow issue so that the political response is more limited than a policy logically derived from its professed goals. While the potential for distorted politics is not a reason to reject any specific proposal for *sui generis* legislation, it is a reason to be wary. This distortion is one reason why *sui generis* rules may ultimately lead to differential treatment of equivalent technologies.

6.6 Weighing It Up

Despite the fact that *sui generis* rules are more likely to be tailored to the specific conduct they deal with than generally framed rules, there are reasons to be cautious in employing them too readily. At the very least, the potential disadvantages of *sui generis* rules should be considered in deciding whether a less well suited, but more general, set of rules might be more appropriate.

Interestingly, these sorts of issues are rarely considered. In the context of the Semiconductor Chip Protection Act of 1984, two congressmen, Robert Kastenmeier and Michael Remington, proposed several factors for enacting *sui generis* intellectual property legislation (1985). They are:

1. The proponent of a new protectable interest ought to show that the interest can fit harmoniously within the existing legal framework without doing violence to existing principles or accepted basic concepts.
2. The proponent of a new intellectual property interest must provide a reasonably clear and functional definition of that interest.
3. The proponent should also provide a valid analysis of the costs and benefits of the proposed legislation on the affected interest groups.
4. The proponent should further show with some specificity how the change will enhance or enrich the public interest.

These criteria are directed at ensuring that legislation is well crafted and properly justified. In a sense, they could be applied to the enactment of *any* legislation – all laws should ideally be well-crafted and properly justified. Kastenmeier and Remington do not refer to any need to consider the special problems that can arise in the context of narrowly framed legislation.

6.7 Tailoring Within Broad Category

There are good reasons to enact *sui generis* rules even where more broadly framed rules could achieve a similar function. As discussed above, broadly framed rules will often be an imperfect fit in a particular context. On the other hand, *sui generis* rules are associated with significant disadvantages – they may be incomplete or expensive, they have a tendency to become obsolete, and they may be designed to favour powerful groups. One way to reduce administrative costs and resolve the problem of incompleteness is to employ a generally based legal regime, but tailor rules within that regime to a particular context.

For example, it is possible to treat human in vitro embryos as potential objects of property, yet create *sui generis* rules to ensure different treatment in some contexts (Bennett Moses 2008). In this case, the rights of control over embryos would be treated as property rights, ensuring the applicability of general laws such as theft and conversion. At the same time, laws applying uniquely to human in vitro embryos could limit the property rights applicable to embryos, in particular by banning certain transactions and restricting permissible conduct to fertilization procedures performed by authorised persons, authorised research and authorised destruction. This would not alter what might be done with embryos, but it would allow property law to deal with intentional and negligent harm. The incompleteness problem is thus solved.

At the same time, the costs of creating rules to deal with the special problems raised by human in vitro embryos are reduced. There is no need to reinvent the wheel and create special rules to deal with every situation where one person's conduct interferes with another's rights to an embryo or damages that embryo. Property rules, made clear through many years of application and interpretation, will deal with those issues. Instead, lawmakers only need to consider the ways in which embryos deserve special treatment. The special respect that embryos deserve due to their potential to become human life can be reflected in specially created *sui generis* rules.

In the intellectual property context, tailoring within a broad category such as patent law is one way to deal with the problem of uniformity cost. Burk and Lemley (2003) propose that *the judiciary* continue to treat different types of inventions differently within the bounds of a broadly framed patent law. Stern (1986) proposes instead that an *agency* be authorised to specify special rules for new technologies within a broadly phrased industrial property system. While delegation has its own problems, it does reduce the risk that specially tailored rules will fall behind the times and then prove difficult to amend (Bennett Moses 2007a).

6.8 Technology Neutral *Sui Generis* Rules

In order to reduce the problem of *sui generis* rules becoming out of date, it is sometimes helpful to draft laws in a technology-neutral way. By this, I mean that a special law can be created to deal with a particular situation, while minimizing the risk that the law will become uncertain, poorly targeted or obsolete in the future. As such, technology-neutral *sui generis* rules will sometimes be a solution to the dilemma presented by Collingridge (1980). Collingridge argued that the social control of technology was difficult because attempts to control a technology early in its development suffer from the difficulty of not knowing its final form and ultimate effects while attempts to control a technology after it had become entrenched were virtually impossible. To the extent that rules designed to deal with specific features of a technology can be crafted in a technology-neutral way, the rules will maintain flexibility as the technology changes.

A technology-neutral *sui generis* approach might be an alternative solution in the case of intellectual property laws such as the Semiconductor Chip Protection Act. Semiconductor chips fell into the "gap" between copyright and patent law because they were functional objects with significant development costs but little in the way of non-obvious innovation. If semiconductor chips deserve intellectual property protection, then it is arguable that *anything* with similar features deserves the same protection. Depending on one's view, semiconductor chips and subpatentable inventions can be protected as part of the law of unfair competition (Janis and Smith 2007), through broadly crafted liability rules (Reichman 1994, 2000) or through a new form of intellectual property.

Like narrowness of legal rules, technology-neutrality is a scale. There are very few goals that can be achieved through perfectly technology-neutral rules that will continue to apply well despite technological change (Bennett Moses 2007a). However, there are drafting techniques that can be used to help make laws more future-proof. In particular, it is possible to employ language that abstracts away from technology-embedded specifics (Bennett Moses 2007a). For example, a "document" (suggesting the use of a physical medium) might become a "preserved communication." While it is possible to enhance the ability of a law to withstand future technological change by careful drafting, there will often be a need to balance technology-neutrality against clarity and operational effectiveness.

A technology-neutral approach is not always appropriate. Koops (2006) gives the example of traffic laws. Such laws commonly distinguish between pedestrians, cyclists and automobiles, thus distinguishing between road-users based on the technology of transportation employed. The need for *sui generis* treatment of bicycles and cars is obvious – the different size and speeds of different vehicles makes different treatment on the road necessary. While it is not necessary to use technology-specific language, it is desirable to do so. One could avoid referring to cyclists specifically by creating rules for those road-users with certain speed and size limitations. Perhaps cycling lanes could only be used by human-propelled vehicles less than three feet wide. But the benefit of such an approach is dubious. While it may help decide which rules apply to futuristic modes of transport, the rules would in the meantime be less clear and could have negative unforeseen effects (Bently 2004 at 176; Reed 2007).

6.9 Conclusion

Especially in the context of technological change, there is a tendency to treat new entities, activities and relationships as in need of special *sui generis* regulation or protection. While legal change is often a necessary response to technological change, it is important to consider the form that any new rules take and, in particular, to bear in mind the costs of *sui generis* rules. In some cases, it may be better to link a new entity, activity or relationship to existing, broadly framed, legal rules while creating exceptions and additions to deal with any special features. It is also worth considering the possibility that *sui generis* rules be drafted with the possibility of future technological change in mind, employing technology-neutral language where there is no significant effect on clarity and ease of application. While this is not an exhaustive account of how to design legal rules that deal with the new conduct made possible by technological change, it hopefully offers some food for thought.

Acknowledgment I am grateful to all participants at the Workshop on Pacing Law and Ethics with Science and Technology for their probing questions and comments. Many of the ideas presented here are derived from the sparks of others, while all errors remain my own.

References

Abelson, H., K. Ledeen, and H. Lewis. 2008. *Blown to bits: Your life, liberty and happiness after the digital explosion*. Upper Saddle River, NJ: Addison-Wesley.

Abramson, B. 2002. Promoting innovation in the software industry: A first principles approach to intellectual property reform. *Boston University Journal of Science and Technology Law* 8: 75–156.

Annas, G., L. Glantz, and P. Roche. 1995. Drafting the Genetic Privacy Act: Science, policy and practical considerations. *Journal of Law, Medicine and Ethics* 23: 360–366.

Bennett Moses, L. 2005. Understanding legal responses to technological change: The example of in vitro fertilization. *Minnesota Journal of Law, Science and Technology* 6: 505–618.

Bennett Moses, L. 2007a. Recurring dilemmas: The law's race to keep up with technological change. *University of Illinois Journal of Law, Technology and Policy* 2007: 239–285.

Bennett Moses, L. 2007b. Why have a theory of law and technological change. *Minnesota Journal of Law, Science and Technology* 8: 589–606.

Bennett Moses, L. 2008. The applicability of property law in new contexts: From cells to cyberspace. *Sydney Law Review* 30(4): 639–662.

Bently, L. 2004. Copyright and the Victorian Internet: Telegraphic property laws in colonial Australia. *Loyola Los Angeles Law Review* 38: 71–176.

Brenner, S. 2007. *Law in an era of "smart" technology.* New York, NY: Oxford University Press.

Breyer, D. et al. 2009. Genetic modification through oligonucleotide-mediated mutagenesis. A GMO regulatory challenge? *Environmental Biosafety Research* 8: 57–64. doi:10.1051/ebr/2009007.

Brownsword, R. 2008. *Rights, regulation and the technological revolution.* Oxford: Oxford University Press.

Burk, D.L. 1991. Biotechnology and patent law: Fitting innovation to the procrustean bed. *Rutgers Computer and Technology Law Journal* 17: 1–85.

Burk, D.L., and M.A. Lemley. 2003. Policy levers in patent law. *Virginia Law Review* 89: 1575–1696.

Callaway, D. 2008. Note: Patent incentives in the semiconductor industry. *Hastings Business Law Journal* 4: 135–151.

Carroll, M.W. 2006. One for all: The problem of uniformity cost in intellectual property law. *American University Law Review*, 55: 845–900.

Collingridge, D. 1980. *The social control of technology*. London: Pinter.

Easterbrook, F.H. 1996. Cyberspace and the law of the horse. *University of Chicago Legal Forum*, 1996: 207–216.

Edgerton, D. 2006. *The shock of the old: Technology and global history since 1900.* London: Profile.

Edwards, R.G., B.D. Bavister, and P.C. Steptoe. 1969. Early stages of fertilization in vitro of human oocytes matured in vitro. *Nature* 221: 632–635.

Ellinson, D. 1988. The patent system – Time to reflect. *Law Institute Journal* 62: 292–293.

Escudero-Pascual, A., and I. Hosein. 2004. The hazards of technology-neutral policy: Questioning lawful access to traffic data. *Communications of ACM* 47 (3): 77–82.

Fitz Simons, J. 1990. Semiconductor chip protection and *sui generis* legislation. In *Essays on Computer Law*, ed. G. Hughes, 51. Melbourne: Longman Professional.

Gostin, L.O., and J. Hodge Jr. 1999. Genetic privacy and the law: An end to genetic exceptionalism. *Jurimetrics* 40: 21–58.

Griem, J.M. Jr. 1993. Note: Against a *sui generis* system of intellectual property for computer software. *Hofstra Law Review* 22: 145–176.

Himma, K. 2003. The relationship between the uniqueness of computer ethics and its independence as a discipline in applied ethics. *Ethics and Information Technology* 5 (4): 225–237.

Hohfeld, W.N. 1913–1914. Some fundamental legal conceptions as applied in judicial reasoning. *Yale Law Journal* 23: 16–59.

Holmes, O.W. 1897. The path of the law. *Harvard Law Review* 10: 457–478.

Janis, M.D., and Smith, S. 2007. The protection of rights in plant varieties: Technological change and the design of plant variety protection regimes. *Chicago-Kent Law Review* 82: 1557–1615.

Johnson, D.G. 2001. *Computer ethics.* 3rd ed.. Upper Saddle River, NJ: Prentice Hall.

Kaplow, L. 1992. Rules versus standards: An economic analysis. *Duke Law Journal* 42: 557–629.

Kastenmeier, R.W., and M.J. Remington. 1985. The Semiconductor Chip Protection Act of 1984: A swamp or firm ground? *Minnesota Law Review* 70: 417–470.
Kirby, M. 2008. New Frontier: Regulating Technology by Law and 'Code'. In *Regulating technologies: Legal futures, regulatory frames and technological fixes*, eds. R. Brownsword, and K. Yeung. Portland: Hart Publishing.
Kohler, P. 2000. The death of ownership and the demise of property. *Current Legal Problems* 53: 237–282.
Kohler, P., and Palmer, N. 1998. Information as property. In *Interests in goods* (2nd ed.), eds. N. Palmer, and E. McKendrick. London: LLP Limited.
Koops, B.-J. 2006. Should ICT regulation be technology-neutral? In *Starting points for ICT regulation: Deconstructing prevalent policy one-liners*, eds. B.-J. Koops et al. (77). National Programme for Information Technology and Law, Information Technology and Law Series No. (9). The Hague: T.M.C. Asser Press.
Kukkonen, C.A. III. 1997. The need to abolish registration for integrated circuit topographies under TRIPS. *IDEA: The Journal of Law and Technology* 38: 105–137.
Lessig, L. 1995. The path of Cyberlaw. *Yale Law Journal* 104: 1743–1755.
Lessig, L. 1999. The law of the horse: What Cyberlaw might teach. *Harvard Law Review* 113: 501–549.
Lewis, T.G. 1995. Comment: Semiconductor chip process protection. *Houston Law Review* 32: 555–613.
Maner, W. 1996. Unique ethical problems in information technology. *Science and Engineering Ethics* (Special Issue: Global Information Ethics, T. Bynum and S. Rogerson eds.) 2 (2): 137–154.
McDonald, J., and K. Christensen. 1998. No jail: Fertility doctor gets home detention, fine. *Orange County Register*, 12 May 1998, B2.
McKeough, J. 1986. Semi conductor chip protection: Copyright or *Sui Generis*? *University of New South Wales Law Journal* 9: 101–116.
Mellor, J. 1988. Patents and genetic engineering – Is it a new problem? *European Intellectual Property Review* 10: 159–162.
Michaelson, P.L. 1986. The 1984 Semiconductor Chip Protection Act – A comprehensive view. *Communications and the Law* 8 (5): 23–55.
Murray, T. 1997. Genetic exceptionalism and "Future Diaries": Is genetic information different from other medical information? In *Genetic secrets: Protecting privacy and confidentiality in the genetic era*, ed. M. Rothstein, 60–73. New Haven, CT: Yale University Press.
Nelkin, D. 1984. Science, technology, and political conflict: Analyzing the issues. *Controversy:* In *The politics of technical decisions,* ed. Dorothy Nelkin, 2nd ed. Beverly Hills, CA: Sage.
Oxford English Dictionary (definition of *sui generis*).
Phillips, J.C. 1992. Note: *Sui generis* intellectual property protection for computer software. *George Washington Law Review* 60: 997–1041.
Purvis, I. 1987. Patents and genetic engineering – Does a new problem need a new solution? *European Intellectual Property Review* 9: 347–348.
Radomsky, L. 2000. Sixteen years after the passage of the U.S. Semiconductor Chip Protection Act: Is international protection working? *Berkeley Technology Law Journal* 15: 1049–1094.
Raskind, L. 1986. The uncertain case for special legislation protecting computer software. *University of Pittsburgh Law Review* 47: 1131–1184.
Rauch, J.G. 1993. The realities of our times: The Semiconductor Chip Protection Act of 1984 and the evolution of the semiconductor industry. *Fordham Entertainment, Media and Intellectual Property Law Forum* 3: 403–439.
Reed, C. 2007. Taking sides on technology neutrality. *Script-ed* 4 (3): 263–284.
Reichman, J.H. 1994. Legal hybrids between the patent and copyright paradigms. *Columbia Law Review* 94: 2432–2558.
Reichman, J.H. 2000. Green tulips and legal Kudzu: Repackaging rights in subpatentable innovation. *Vanderbilt Law Review* 53: 1743–1798.

Risberg, R.L. Jr. 1990 Comment: Five years without infringement litigation under the Semiconductor Chip Protection Act: Unmasking the spectre of chip piracy in an era of diverse and incompatible process technologies. *Wisconsin Law Review* 24: 241–277.
Samuels, L.B., and J.M. Samuels. 1986. Semiconductor Chip Protection Act of 1984: An analytical commentary. *American Business Law Journal* 23(4): 601–616.
Samuelson, P. 1985. Creating a new kind of intellectual property: Applying the lessons of the chip law to computer programs. *Minnesota Law Review* 70: 471–531.
Samuelson, P. et al. 1994. Manifesto concerning the legal protection of computer programs. *Columbia Law Review* 94: 2308–2431.
Schlag, P. 1985. Rules and standards. *University of California Los Angeles Law Review* 33: 379–430.
Shi, Q. 2005. Patent system meets new sciences: Is the law responsive to changing technologies and industries? *New York University Annual Survey of American Law* 61: 317–347.
Sommer, J.H. 2000. Against Cyberlaw. *Berkeley Technology Law Journal* 15: 1145–1232.
Stern, R.H. 1986. The future of software protection: The bundle of rights suited to new technology. *University of Pittsburgh Law Review* 47: 1229–1267.
Svantesson, D.J.B. 2007. *Private International Law and the Internet*. Alphen aan den Rijn: Kluwer Law International.
van der Haar, I. 2007. Technology neutrality; What does it entail. *TILEC Discussion Paper, DP2007-009*.
Wahlgren, P. 2004. IT and legislative development. *IT Law, Scandinavian Studies in Law* 47: 601–618.
Weber, T., and J. Marquis. 1995. In quest for miracles, did fertility clinic go too far? *LA Times*, 4 June 1995, A1.
Williams, J.J. 2005. Protecting the frontiers of biotechnology beyond the genome: The limits of patent law in the face of the proteomics revolution. *Vanderbilt Law Review* 58: 955–994.
Wilson, J.Q. 1980. The politics of regulation. In *The politics of regulation*, ed. J.Q. Wilson, 357–394. New York, NY: Basic Books.

Legislation

Semiconductor Chip Protection Act, 17 U.S.C. 901-914
Telephone Consumer Protection Act, 47 U.S.C. 227
CAN-SPAM Act, 15 U.S.C. 7701-7713
Californian Penal Code, §§ 367g, 503

Cases

Dahl v. Angle, 222 Ore. App. 572.
Davis v. Davis, 842 S.W.2d 588 (Tenn., 1992).
Del Zio v. The Presbyterian Hospital in New York, No. 74 Civ. 3588, 1978 US Dist LEXIS 14450 (SDNY, 1978).
Frisina v. Women and Infants Hospital, No CIV A 95-4037, No CIV A 95-4468, No CIV A 95-5827, 2002 WL 1288784 (Sup. Ct. R.I., 30 May 2002).
Jeter v. Mayo Clinic Arizona, 121 P.3d 1256 (Ct. App. Ariz., 2005).
J.E.M. Ag Supply, Inc. v Pioneer Hi-Bred Int'l, Inc. 534 U.S. 124 (2001).
York v. Jones, 717 F. Supp. 421 (E.D. Va., 1989).

Other

Australian Law Reform Commission. 2003. Report 96, *Essentially Yours: Protection of Human Genetic Information in Australia.*

H.R. Rep. No 781, 98th Cong., 2d Sess., reprinted in 1984 US Code Cong & AD News 5750.

Intellectual Property and Competition Review Committee. 2000. *Review of Intellectual Property Legislation under the Competition Principles Agreement* available at http://www.ipaustralia.gov.au/pdfs/ipcr/finalreport.pdf. (Sept 2000).

Chapter 7
Anticipatory Governance of Emerging Technologies

Daniel Sarewitz

In the last two centuries continual technological innovation has catalyzed wave after wave of social transformation, and is implicated in a few cataclysms as well. New waves now seem, perhaps, to be rapidly approaching, associated with accumulating and converging scientific and technological advances in such areas as nanotechnology, information technology, neuroscience, and human biotechnology. The power of these emerging technologies to remake society is thought by many to be on a scale comparable to the rise of steam power in the first industrial revolution, the emergence and convergence of electric power and the internal combustion engine in the late nineteenth century, and the proliferation of information technologies in the latter part of the twentieth century. Energy production systems, manufacturing systems, military weaponry, even the performance standards of the human brain and body are seen by some as subject to radical transformation in the coming decades. The accuracy of particular technological predictions is not really important. What is undeniable is that the scale and pace of the global research and innovation effort continues to grow, and that the consequences of this effort continue to permeate and transform society at every level.

What are the prospects for governing the societal implications and consequences of these emerging waves of technology? Current approaches are almost entirely reactive, ponderous, and bureaucratic, and are increasingly overmatched by the scale and pace of technological innovation and change. Standard regulatory regimes for dealing with chemicals in the environment, for example, have devolved into a miasma of litigation, politics, and scientific uncertainty that benefits neither the environment nor the economy. Governance of pharmaceutical products is perhaps justly criticized from all sides – useful drugs are not approved fast enough, harmful ones are not caught soon enough, useless ones seem to proliferate. Innovation is simply too fast, too pervasive, too decentralized to yield to approaches that demand comprehensive knowledge as a basis for taking action.

D. Sarewitz (✉)
Arizona State University, Tempe, AZ, USA
e-mail: daniel.sarewitz@asu.edu

G.E. Marchant et al. (eds.), *The Growing Gap Between Emerging Technologies and Legal-Ethical Oversight*, The International Library of Ethics, Law and Technology 7,
DOI 10.1007/978-94-007-1356-7_7,

Is there a way forward? I start with some general thoughts about what it actually means to talk about governing technological change, before moving toward some concrete examples of work now being done to develop theories, models, and tools that can improve the social capacity for guiding future technologies toward desired societal outcomes and away from undesired ones, a process termed anticipatory governance (Barben et al. 2008).

The dilemma for democratic societies created by our commitment to continual technological advance is obvious. If on the one hand we are committed to notions of pluralism, participation, and openness in charting the course of society, how on the other can we come to grips with the enormous transformational power of technology and technological systems, a power that often seems at once inscrutable, unconscious, overwhelming, and autonomous? Thirty years ago Langdon Winner (1978) developed the notion of "reverse adaptation" to describe the "adjustment of human ends to match the character of the available means" created by technological systems. At around the same time, David Collingridge (1980) articulated the fundamental dilemma of technological governance: in the early phases of technological evolution, many avenues of advance are available and promising, but too little is known about potential impacts to choose the best paths. Later on, when more is known about impacts, options are greatly restricted due to technological lock-in and concentration of power among vested interests. In light of such observations, and given that technology is among the most powerful forces for social transformation operating in the world today, it's not unreasonable to wonder about the extent to which our commitment to democracy is an illusion or an opiate.

Of course one could say the same thing about, say, earthquakes, weather, or the motion of the solar system, that they make a mockery of democratic aspirations since they mediate our actions without our consent. But no one complains that the laws of gravity, or the motion of tectonic plates, are unfair and need to be governed more wisely. So we similarly could – and often do – place technological change outside of ourselves by conceiving it as an external phenomenon. This solves the democracy problem, because it allows us to treat technological change as a force to which we can only react. And indeed, for the most part, our approach is to pour tremendous resources into the creation of technological advance and then regulate and react to the outcomes as necessary to make them tolerable, just as we react to and accommodate weather or the motions of the Earth's crust.

But this remains unsatisfying because technological innovation is, after all, a human endeavor, one that arises from human choices made in human cultures and institutions, and one whose importance for society depends on the continual willingness of humans to avidly make use of technological products and processes. Technology is our unruly child and we cannot evade some sense of accountability for its behavior.

Starting in the late 1960s, in the shadow of the Cold War and the emergence of the environmental movement, aspirations for the control of technological change began to grow. In particular, the technology assessment movement was rooted in the notion that future trajectories of technological evolution could be predicted and governed. As explained in 1976 by Harvey Brooks (p. 20), one of the founders of the study of post-War science policy: "Ideally the concept of TA [Technology Assessment] is that it should forecast, at least on a probabilistic basis, the full spectrum of possible consequences of technological advance, leaving to the political process the actual choice among the alternative policies in the light of the best available knowledge of their likely consequences."

But a more pessimistic vein of analysis, represented by people like Winner, Collingridge, and, before them, Lewis Mumford and Jacques Ellul, viewed such hyper-rational ambitions as implausible, due to the pervasive embeddedness of technological innovation in human institutions and political arrangements. Winner (1978) talked about "technics out-of-control" and "technological somnambulance" to convey a sense of powerlessness and resignation in the face of continual technological transformation of society.

What lies between an implausible commitment to control and a fatalistic embracing of passivity? Certainly the expectation that democratic societies (or any other societies, for that matter) can dictate technological futures is neither coherent nor desirable. We know that efforts to control most forms of social activity turn out to create more problems than they prevent. And we know that most efforts to predict technological pathways as an input into decision-making have been failures, and often absurd failures at that. But neither is there a need to abandon all hope. Another way to look at the problem is to start with the recognition that, like procreation, technological innovation is an innately human activity, and as such it acts as a mirror on, and amplifier of, the ambiguities and contradictions of the human condition. If this recognition tempers our expectations, then the alternative to control is not abdication, but reflection – on what we are actually doing – and governance – based on our reflections, and carried out in the context of our democratic aspirations. In most other domains of important human action and choice, the role of democratic decision making is not to exercise control but to reveal and adjudicate value disputes that underlie choices. Yet it is precisely in this domain that governance of technological change, for the most part, has gotten a free pass.

Why should this be? The key reality is that the products of science and technology do not appear magically; rather, they emerge from choices made by people working in institutions designed by people. In the United States after World War II, a series of strategic decisions were made about which areas of science should be advanced, and those decisions led, over a period of several decades, to revolutions in such areas as computer science, solid-state physics, materials science, molecular biology, genomics, and electrical engineering, and to linked technological revolutions in weapons, communication, information, transportation, and bio-technologies. These developments were not designed in advance and implemented in an ordered or predictably way. But neither did they happen accidentally, serendipitously, randomly, surprisingly. It was all a product of decisions made in government, in industry, in universities, by people with a strong, if evolving, sense of what they were trying to accomplish over the long term. The process was powerfully driven by the role of the U.S. Department of Defense as both leading investor in, and principal consumer of, advanced technology (Alic John 2007). Yet the approach was dominated not by top-down planning, but by catalyzing close relationships among a relatively small number of leading universities, corporations, and government agencies. These linkages led to tightly coupled networks and feedbacks across a growing innovation enterprise that was at once institutionally highly complex, yet highly focused, as a matter of mission, on rapid technological advance. In other words, the explosive growth of the U.S. R&D enterprise in the Cold War era was an exercise in the governance of science and technology within the broader context of a complex, adaptive innovation system. Similarly, any effort to govern the societal implications of rapidly emerging technologies must contend with, and indeed exploit, the decentralized, networked essence of the innovation process.

The systems view renders standard cause-effect thinking irrelevant. For example, the question of whether the long-term results of some specific discovery or line of research actually were predictable was quite besides the point. Decisions were being made with a view toward future outcomes, not by tossing dice, and such decisions strongly determined what types of knowledge and innovation were created, and who was likely to benefit from that knowledge and innovation. Decision makers were acting in response to values, interests, aspiration, power, etc., just as decision makers always do. *The key questions, then, are these: who is making the decisions? And how do these decisions emerge from and interact with the complex socio-technical context within which they are being made?*

Why has technological change, unlike other areas of human activity, largely been exempted from the rigors of democratic debate? Certainly part of the reason, as I've suggested, is the sense that technological change is simply too complicated, too unpredictable, and too inevitable, to yield to collective engagement with its meanings in democratic forums. Another reason is that technology is closely aligned with science, and so with the powerful cultural belief that the process works best when it is left alone. A related supporting belief is that benefits are inherent in science and technology, whereas the problems they create are the fault of society, or politics. Perhaps most importantly, however, is the alignment of technological innovation with the ideologies of the marketplace, which tell us that the appropriate measures of technological value are monetary, and the appropriate mode of intervention is hands-off.

Whatever hypothesis one prefers, the overriding fact is that, in contrast to almost every other important area of human endeavor, the pursuit of technological transformation is largely exempted from formal democratic processes of eliciting value preferences and adjudicating value disputes about desired future states, even though technological innovation strongly expresses those very things.

This exemption perhaps explains why Technology Assessment began as a technocratic exercise, with rational analysis formally separated from political decision processes. TA was something added on to the innovation process, done in different places, like the defunct Congressional Office of Technology Assessment. TA also bought into the notion of science and technology as essentially autonomous enterprises that could be governed by introducing new technical information into political discourse as a basis for regulation. Thus, TA harbored the expectation that decision makers would potentially be willing to make controversial decisions on the basis of highly contestable, non-verifiable probabilistic statements about the future of a technology. It was destined to disappoint. As Harvey Brooks (1976, p. 21) wrote: "The record on the implementation of TA has not been particularly happy. The outcome, whether negative or positive, tends to be more determined by political momentum and bureaucratic balance of power than by a rational process."

Surprise. But this then led to the wrong conclusion: that, since technological assessment could not be based on technocratic predictions, it could not be done at all. This wrong conclusion was inherited by the next generation of technology governance through the Ethical, Legal, and Societal Implications (ELSI) program of the Human Genome Project. In the early 1990s, ELSI was grafted onto the Genome Project to support research by social scientists and humanists on some of the complex dilemmas raised by the coming proliferation of genomic information (Cook-Deegan 1996). ELSI was about understanding emerging social dilemmas, but it included no mechanisms for feeding back into decision making about science, or feeding forward into decisions about genome politics. It codified the separation of the science from the study of the social outcomes of science, and marks the end of the first era of Technology Assessment.

In contrast, over the past several decades, growing insight into the dynamics of innovation systems has stimulated new approaches to technological governance aimed at resolving the Collingridge dilemma. These new approaches are rooted in the idea that, by making the human choice contexts implicated in innovation processes visible and open to multiple perspectives, conscious governance can emerge at earlier stages, when more options are available and when uncertainty about future impacts is higher. This work was pioneered by Arie Rip and colleagues in the Netherlands, who termed it "constructive technology assessment" (Schot and Rip 1997) and has more recently gained beachheads in Britain, for example with work done at Lancaster University and the think-tank Demos on "upstream engagement" (Wilsdon and Willis 2004), and in the U.S., for example with work I've been involved with at Arizona State University (ASU), which we term "real-time technology assessment" (Guston and Sarewitz 2002), and which I will describe in more detail below. The goal of these efforts, most broadly, is to inject pluralistic reflection into the innovation process as a means of improving the public value of new technologies. The overall goal is to create a capacity for anticipatory governance, by building reflexivity into institutions where key technoscientific choices are being made.

If we understand that we are all participants in a great experiment in social transformation being carried out without our consent or even our understanding, the self-imposed limits of TA now become almost painfully obvious. If we understand technological transformation as emerging not from the autonomous, automatic advance of science and technology but from a complex set of decisions made within a variety of institution contexts, then a different way to think about and implement TA can emerge. This new approach to TA will reflect the following realities:

1. The pace and direction of advancing knowledge and applications is determined by human choice.

2. The specific directions that technoscience is steered, and the pace of its advance, reflect who is making the decisions – their interests, values, motives, perspectives.
3. The decisions that are made are determined within a complex social setting that encompasses a range of socioeconomic, cultural, and political components.
4. This complex social setting interacts with the results of technoscientific advance to yield social outcomes. The setting, the science, and the outcomes mutually evolve over time.

These realities raise the following questions:

1. What is the range of choices available to people making decisions about science?
2. What are the interests, values, motives, and perspectives of people making decisions about science?
3. How do these interests, values, motives, and perspectives relate to the complex social setting within which decisions are made?
4. How do the results of scientific advance interact with socioeconomic, cultural, and political factors to yield social outcomes?

These questions can be researched and understood to various extents and would constitute *both the intellectual and the operational agenda* for the new approach to technological governance – an anticipatory approach, not in the futile sense of first predict, then take action, but in the sense of building institutional capacities to reflect on contexts and choices. The goal is to build a capacity for *reflexiveness* – social learning that expands the realm of conscious and available choice – into science and technology institutions and decision processes themselves. The process of understanding the dynamics of decision making about science and technology simultaneously provides knowledge and insight that can improve decision making processes and enable the participation of a broader and more diverse community of decision makers. Decision making is improved because previously implicit decisions become explicit, because expanded choices relevant to the decisions become apparent, and because greater diversity of actors relevant to the decisions can recognize themselves as potential stakeholders, thus creating the potential for improved deliberation.

While this capacity for reflexivity and anticipatory governance can and should be enhanced at many points in the innovation system, work that I've been involved in at Arizona State University focuses on the very upstream end of an emerging class of technology – nanotechnology – in the laboratory setting itself. The Center for Nanotechnology in Society at ASU (CNS; cns.asu.edu), funded by the National Science Foundation, is in essence a test-bed for the idea that reflexivity can be built into the research process via a suite of social science methods termed by that constitute "real-time technology assessment," or RTTA (Guston and Sarewitz 2002).

Above all else, RTTA, is about institutional innovation. It's about taking the closed environment of the research institution and opening it up so that the complex social dynamics of early-stage innovation processes become apparent to those who are most centrally involved in those processes. RTTA includes four activities that, when taken together, create what are intended to be the necessary components of an inherently reflexive research process. The four components are:

- RTTA 1: Innovation system analysis. This activity builds knowledge of the technical landscape and the opportunities it is enabling.
- RTTA 2: Analysis of scientists' and the public's values and attitudes. This creates a generous awareness of the diverse values and aspirations of current and potential stakeholders who inhabit that landscape.
- RTTA 3: The creation of abundant opportunities for deliberation and stakeholder participation, informed and structured by what we learn in RTTA 1 and 2. This allows expansive exploration of alternative potential futures and landscapes.
- RTTA 4: Assessment of the social learning that actually occurs as a result of the prior three activities. This activity builds empirically grounded insight into how the system itself is evolving, and feeds that understanding back into the system.

RTTA seeks to make conscious and explicit the complex social, political, and economic setting within which nanotechnology research and innovation occurs, as it is occurring. RTTA does not try to predict the future of nanotechnology, but it does aim at stimulating discussions about what types of futures are possible, and what types are desirable. RTTA is certainly not in the business of telling researchers what to do, but it is in the business of allowing them to understand what they are doing in a manner that is much more contextually rich than in usual laboratory settings.

As I have emphasized, institutional innovation is at the heart of the effort. The goal is to move toward a research setting that is highly permeable to ideas and concerns that are normally excluded from lab settings. It took much of CNS's first 2 years simply to put the collaborative networks in place, build the necessary trust among partners, and begin to fully implement the wide array of opportunities for reflexive engagement, including collaborative teaching, joint research activities, and informal discussions between nanoscientists and social scientists; science cafes and other events in the community; scenario workshops and other future-visioning activities; and shared support of graduate students.

In March 2008, as the most important of CNS's participatory, RTTA 3 activities, we held the first National Citizen's Technology Forum (NCTF), bringing lay citizens together at six sites across the national to discuss, in highly mediated settings, the social implications of rapidly emerging and converging technologies. (For a full description, see Hamlett et al. 2008.) Some of the major issues of interest that emerged included:

- Need for effective regulation of emerging technologies;
- Demand for effective programs of public information;
- Concern about equitable and needs-based access to new technologies;
- Ambivalence about privacy, safety, and human performance enhancement.

Crucially, the deliberative process itself led to shifts in attitudes over the month-long course of the NCTF – that is to say, people actually did deliberate. While participants were almost uniformly optimistic about emerging technologies both before and after the NCTF, concerns about the downsides increased markedly. Doubts about the benefits of applications for human enhancement, about equitable access to technologies, about risks, and about economic implications all increased as a consequence of the extended deliberative process. That is to say, the deliberations enhanced the intellectual sophistication of the participants by allowing them to hold internally conflicting views of nanotechnology. This outcome is encouraging as well, but it also predicts that technical communities may seek to advance their own interests by opposing efforts to expand RTTA-like institutional innovations in the R&D system.

Within the laboratory setting itself, RTTA is intended to enhance awareness of the choices that researchers face as they pursue their experiments. Several types of questions seem to be surfacing at CNS, for example:

- Given several research project options, which one is likely to yield the most social benefit in the near term?
- Given several molecules that can serve a particular function, which one is the most environmentally benign?
- Should a neural enhancement device be implanted in the brain or be worn externally?
- Are the potential benefits of a human memory-enhancement implant obviously going to be greater than the potential downsides?

Yet such questions are perhaps overly concrete, because they might seem to suggest that the idea is to directly link individual choices upstream in the laboratory to complex downstream consequences in society. Innovation system complexity means that cause-effect chains will always be difficult to trace and that, except in exceptionally rare cases, the consequences of individual decisions will not be discernible in broad societal outcomes. Rather RTTA is a tool to build systemic capacity – the capacity to reflect on context and choice at a multitude of times and places in the innovation process.

At CNS, one early place where we expect to see evidence of this enhanced capacity is in the evolution of values and attitudes of scientists, engineers, and social science researchers to reflect greater awareness of the political, social, and economic contexts of innovation. We would also expect to see institutional values and norms evolve as well, for example in terms of expanded notions of scientific responsibility and productivity, and of what successful graduate education should look like. These are hypothesis that are still being tested as CNS moves into its fifth full year of activity.

The overarching hypothesis behind CNS and RTTA is that an emerging reflexive capacity will favor more socially beneficial choices – that is, choices that steer toward articulated public values – and CNS researchers are continually testing this hypothesis at the micro-level of partner research laboratories. Indeed, CNS is motivated by the belief that the very process of turning laboratories at a research university from insular to openly reflexive is inherently beneficial because it creates openness, transparency, and broader capacity for engaged deliberation than existed previously. This benefit is in part procedural, in that open and aware deliberation is more democratically satisfactory than closed and clueless deliberation, or than a lack of conscious deliberation. But CNS also tests the idea that the benefit is instrumental: that reflexivity moves innovation toward more socially desirable outcomes, and away from undesirable ones, as diverse decision makers reflect more deeply on the context of their decisions. And of course this can happen either through a change in innovation paths, or through a change in the conceptions of desirability, or, more likely, through the interaction of both.

This conscious yet non-deterministic evolutionary process is at the heart of anticipatory governance. Anticipatory governance is an appropriate aspiration for democratic engagement with technological transformation, one that succumbs neither to the illusion of control, nor the resignation of technological somnambulance.

Anticipatory governance comprises three areas of simultaneous activity: engagement, foresight, and integration. Engagement encompasses the suite of activities that stimulates public deliberation; foresight describes the process of developing plausible and evolving scenarios of possible futures that can be the subject of the public deliberation; and integration brings the engagement and foresight activities into the domain of scientific practice to enhance reflexiveness (Barben et al. 2008).

RTTA represents one suite of methods aimed at advancing the goal of anticipatory governance, at one site, at one university. As I have mentioned, there are a few other similar exercises taking place, mostly in western Europe. If we are to escape from our self-imposed subjugation at the hands of the Collingridge dilemma, then the challenge is to move from local experiments and pilot projects to a scaled up, society-wide capacity to innovate reflexively, rather than unconsciously.

Part of the challenge is simply to make is it safe to talk about innovation in terms of a range of public values and choices, rather than in the simple input–output, more-is-always-better mode. For example, it's not hard to think of some fairly simple questions that could *always* be discussed in public venues when decisions are being made about what R&D will be done. Instead of just asking: How much should we spend on this program or that? We can also ask:

- What are the values that motivate a particular investment in innovation?
- Who holds those values?
- Who is most likely to benefit from the translation of the research results into social outcomes? Who is unlikely to benefit?
- What alternative approaches are available for pursuing such goals?
- Who might be more likely to benefit from choosing alternative approaches?
- Who might be less likely to benefit?

The habit of asking these sorts of questions has not yet been formed. But habits do change. Important norms of scientific practice, for example, have evolved greatly in the past several decades. Issues of human subjects research, of the use and treatment of animals in research, of environmentally safe practice, of the gender and ethnic diversity of the scientific community, have all become mainstream concerns of policy makers and researchers alike, whereas in the recent past, serious consideration of such issues was often labeled as "anti-scientific."

Moreover, these changes in norms have come about along with changes in institutional structure. For example, concern about the ethical governance of human subjects research in the U.S. has led to nationwide institutional reform. Every publicly funded research project involving human subjects is monitored by an institutional review board (IRB) that must approve the research before it can be conducted, and ensure that ethical principles such as prior informed consent are enforced. There are thousands of such boards operating in the United States, thus demonstrating that comprehensive governance of innovation activities is a reasonable goal. While IRBs are far from perfect in protecting the rights of research subjects, and while they also impose a cost in terms of the efficiency of conducting research, they are nonetheless an accepted element of a scientific infrastructure that respects and protects human dignity.

The IRB experience demonstrates that comprehensiveness is possible when the stakes are high – and the stakes associated with emerging and converging technological revolutions are enormous and radical. Just as the IRB process is an accepted part of *all* human subjects research, institutionalizing anticipatory governance activities as part of the publicly funded science and technology enterprise could be done by requiring an RTTA-like component for *all* major public programs and projects related to transformational technoscience. This capacity-building could be funded by a small tithe, perhaps 2%, on research and innovation expenditures. And while such a scenario may seem, right now, to be ridiculously ambitious, one could easily imagine a time, perhaps several decades in the future, when every major research institution would be continuously engaged in the process of reflecting upon the values and choices that are implicated in its work. At such a time in the future, what will truly seem ridiculous is the fact that major research institutions in the first decades of the twenty-first century were committed to a rejection of the need for continually reflecting on the social meanings of the emerging technologies that they help to create.

References

Alic, John. 2007. *Trillions for military technology: How the pentagon innovates and why it costs so much*. New York: Palgrave Macmillan.

Barben, D., E. Fisher, C. Selin, and D. Guston. 2008. Anticipatory governance of nanotechnology: Foresight, engagement, and integration. In *The handbook of science and technology studies*, eds. E. Hackett, O. Amsterdamska, M. Lynch, and J. Wajcman, ch. 38, 979–2000. Cambridge, MA: MIT Press.

Brooks, Harvey. 1976. Technology assessment in retrospect. *Newsletter on Science, Technology, and Human Values*, no. 17 (Oct).

Collingridge, D. 1980. *The social control of technology*. London: Frances Pinter.

Cook-Deegan, R. 1996. *The gene wars: Science, politics, and the human genome project*. New York: W.W. Norton.

Guston, D., and D. Sarewitz. 2002. Real-time technology assessment. *Technology in Society* 24: 93–109.

Hamlett, P., M. Cobb, and D. Guston. 2008. *National citizens technology forum: Nanotechnologies and human enhancement*. CNS-ASU Report #R08-0003. Tempe, AZ: Center for Nanotechnology in Society, Arizona State University.

Schot, J., and A. Rip. 1997. The past and future of constructive technology assessment. *Technological Forecasting and Social Change* 54 (2–3): 251–268, Feb Mar 1997.

Wilsdon, J., and R. Willis. 2004. *See-through science: Why public engagement needs to move upstream*. London: Demos.

Winner, L. 1978. *Autonomous technology: Technics-out-of-control as a theme in political thought*. Cambridge, MA: MIT Press.

Part III
A Toolbox of Solutions

Chapter 8
Pacing Science and Technology with Codes of Conduct: Rethinking What Works

Brian Rappert

8.1 Introduction

Against social, political, and ethical concerns associated with developments in science and technology (S&T), continuing suggestions have been forwarded that scientists and engineers should adopt what are generically referred to as "codes of conduct."[1]

By way of understanding the utility of codes in addressing societal challenges, this chapter provides an in-depth analysis of one set of initiatives: attempts to prevent the destructive use of life science research findings and techniques. Particularly since 9/11 and the US anthrax attacks, concerns have been raised by a diverse range of organizations about whether the potential for life science fields to transform health and research techniques might facilitate the deliberate spread of disease; and if so, what responsive measure should follow to minimize these threats. For instance, reports such as the 2003 US National Academies' *Biotechnology Research in an Age of Terrorism,* the 2006 Royal Society-InterAcademy Panel-International Council for Science's *Scientific and Technological Developments Relevant to the Biological & Toxin Weapons Convention*, and the 2008 report *World at Risk* from the US Commission on the Prevention of Weapons of Mass Destruction Proliferation and Terrorism have suggested how developments in civilian research across a wide range of disciplines are helping to lower the barriers to and enhance the power of bioweapons.[2]

---

[1]Examples of code initiatives in nanotechnology include, for instance, the European Commission's 2008 Recommendation document *On a Code of Conduct for Responsible Nanosciences and Nanotechnologies Research*, the European Nanotechnology Trade Alliance's *Developing a Nanotechnology Code of Conduct for European Industry*, the collaborative *Responsible NanoCode*.

[2]Royal Society. 2006. *Report of the RS-IAP-ICSU international workshop on science and technology developments relevant to the Biological and Toxin Weapons Convention.* London: Royal Society.

B. Rappert (✉)
University of Exeter, Exeter, Devon, UK
e-mail: B.Rappert@exeter.ac.uk

G.E. Marchant et al. (eds.), *The Growing Gap Between Emerging Technologies and Legal-Ethical Oversight*, The International Library of Ethics, Law and Technology 7,
DOI 10.1007/978-94-007-1356-7_8,

In this way, the concern is that the knowledge derived from life science fields – such as virology, molecular genetics, neuroscience, synthetic biology and elsewhere – could *further* rather than *prevent* the spread of disease. As the potential accorded to the life sciences to revolutionize our understanding of the world for benefit are stressed in popular and policy discussions, many have identified security concerns that follow on since post-9/11. With such profound and wide ranging concerns, challenging questions have been asked about what sort of research should be done, under what conditions, and whether it should be communicated.[3]

A widespread starting presumption of these discussions has been that the said ever accelerating pace of developments and (related to this) the worldwide proliferation of life science research frustrate traditional approaches of devising formal rules and regulations. As elsewhere, but perhaps particularly pronounced in this case, "codes of conduct" have been forwarded as means of (largely) self-regulation adept enough to keep pace with the speed of developments in S&T. Codes have received significant (renewed) attention in numerous national and international settings as ways of fostering a "culture of responsibility," most notably during the 2005 and 2008 meetings of governments to the Biological and Toxin Weapons Convention (BWC). Since 2003, the author has observed and actively contributed to varied efforts along these lines.[4]

Beyond recounting one area where codes have been offered as a way to address a perceived gap between accelerating technology and lagging regulatory oversight, this paper seeks to reframe traditional approaches to codes' utility more generally. This will be done by not just asking whether bioweapons-related codes have "worked" in plugging gaps, but by questioning what "working" should be taken to mean. While codes have been portrayed as instruments to guide the behavior of practitioners, many factors have frustrated achieving this aim. As will be contended, it is in the very process of deliberating about codes that codes have had most significance in helping track S&T – such as through building of shared agendas and enabling future co-ordinated initiatives. In many respects, the outcome of codes talk has been their enactment. The limits and dilemmas of this role will lead to a consideration of the place of skepticism and belief in the policy making process.

8.2 Some Preliminary Points

Although "codes of conduct" are hardly novel, in recent decades they have increasing been offered as means of responding to and avoiding dubious practices across

[3]Rappert, Brian. 2008. Defining the emerging concern with biosecurity. *Japan Journal for Science, Technology and Society* 17: 95–116.

[4]Facilitated by a grant from by the UK Economic and Social Research Council (ESRC) New Security Challenges Program (RES-223-25-0053) running from 2004 to 2007. See Rappert, Brian. 2009. *Experimental secrets.* Lanham, NY: University Press of America.

many domains.[5,6] This term is routinely used to refer to activities with a wide range of:

- aims (for instance, aspire, educate, foster ethical debate, prescribe or proscribe behavior)[7];
- drafters (for instance, professional associations, companies and other organizations, governments, inter-organizational bodies); and
- target audiences (for instance, individuals, professional bodies, members of industrial alliances).[8]

As a result, what is labeled as a code of conduct with regard to one set of issues (for instance, corporate environmental responsibility) might be quite different in character than one for another area or even other codes addressing the same set of issues. With this elasticity, the scope for misunderstanding and cross-talk is considerable.[9]

Historically in the biosciences, efforts to devise codes have been less intense than in engineering or medicine where professional-client relations have raised recurring questions about appropriate conduct.

Social scientists and ethicists who have examined codes in science and engineering typically have done so through asking two questions:

> Do codes work?
> And, could codes work?

With regard to the former, contrasting claims are often made about the utility of science codes.[10] Aspirational, educational, and advisory orientated ones have been criticized for being vague, open to multiple interpretations, ineffective to stop would-be trespassers, and often poorly known. As well, it has been argued that the provisions of codes tend to codify existing practices rather than set new standards that could change behavior.[11] Others have rejoined that rather than being a way to change behavior, they can help raise awareness about important topics, alert individuals to specific sensitive matters, foster standards and ethical reflection about

---

[5]See www.codesofconduct.org for many written examples.

[6]Kaptein, Muel. 2004. Business codes of multinational firms. *Journal of Business Ethics* 50: 13–31.

[7]Rappert, B. 2004. Responsibility in the life sciences. *Biosecurity and Bioterrorism* 2(3): 164–175.

[8]Royal Netherlands Academy of Arts and Sciences. 2007. *A code of conduct for biosecurity*. Amsterdam: KNAW.

[9]While a matter of speculation based on my personal experience, the uncertainty about what is meant by the term 'code of conduct' is probably highly functional in contributing to suggestions of their utility.

[10]Rappert, Brian. 2004. *Towards a Life Sciences Code: Countering the Threats from Biological Weapons*. Bradford Briefing Papers (2nd series); No. 13 See http://www.brad.ac.uk/acad/sbtwc/briefing/bw-briefing.htm

[11]Pels, P. 1999. Professions of Duplexity. *Current Anthropology* 40(2): 101–114.

emerging issues, clarify responsibilities, and increase public confidence.[12] Much of the evaluation of the sub-set of codes that include sanctionable requirements turns on the case by case commitment made to their enforcement.

With regard to the question of whether codes *could* work, some have argued that to think abstract guidelines could determine appropriate conduct for specific situations misconstrues the nature of ethical decisions.[13]

8.3 Codes and Biological Weapons: Expectations and Transformations

The potential range of initiatives that can be labeled as a "code" and of criteria that might be brought to bear in their evaluation make it rather restrictive to define in advance what counts as an instance of one. Rather, in considering their utility for pacing S&T, being open to how codes are multiply defined allows for an appreciation of the diversity of agendas being sought.

In relation to concerns associated with the destructive potential of life science research, an indication of the range of types, purposes, and changing expectations for the codes suggested is indicated by Box 8.1.

Box 8.1 Proposals for Biosecurity Codes

A Hippocratic Oath for Scientists?

Proposal: As part of attention to terrorist threats after 9/11, a 2003 report by the British House of Commons Science and Technology Committee titled *Scientific Response to Terrorism* suggested that "an overt ethical code of conduct linked to professional membership analogous to the Hippocratic Oath" be established for those working with dangerous substances or pathogens.[14] The Committee further added that if "the scientific community does not take stronger action to regulate itself then it risks having ill-judged restrictions placed on it by politicians."[15]

---

[12]See as well Atlas, R., and M. Somerville. 2007. Life sciences or death sciences. In *Web of prevention*, eds. B. Rappert and C. McLeish. London: Earthscan.

[13]Ladd, J. 1991. The question for a code of professional ethics. In *Ethical issues in engineering*, ed. D. Johnson, 130–136. Upper Saddle River, NJ: Prentice Hall.

[14]House of Commons Science and Technology Select Committee. 2003. *The scientific response to terrorism*. HC 415-II, Examination of Witnesses, May 14, 2003. London: HMSO.

[15]Ibid.: paragraph 211. See as well Times Higher Education Supplement. 2003. Agree ethics code or face state control. *Times Higher Education Supplement*, 14 Nov.

Result: While this suggestion was directed at learned and professional societies as well as public research funding agencies, to have linked a code to membership would have required introducing a new framework for controlling who can practice research (since, for instance, membership of a scientific society is rarely needed to conduct research). Entry into science is not licensed in the same way as fields such as engineering and medicine in the UK. No such membership ethical code was introduced along the lines the Committee proposed – either by scientists or politicians. No stronger impositions followed as warned.

Uniting Around a Restricting Code?

Proposal: In 2002, the Working Group of the United Nations and Terrorism recommended that "Relevant United Nations offices should be tasked with producing proposals to reinforce ethical norms, and the creation of codes of conduct for scientists, through international and national scientific societies and institutions. . .[s]uch codes of conduct would aim to prevent the involvement of defence scientists or technical experts in terrorist activities and restrict public access to knowledge and expertise on the development, production, stockpiling and use of weapons of mass destruction or related technologies".[16]

Developments: The International Centre for Genetic Engineering and Biotechnology (ICGEB) – a UN provider of training in biotechnology for countries of the developing world – was tasked by the UN Assistant Secretary for Disarmament with this responsibility. The ICGEB sought to collaborate with the Inter-Academy Panel (IAP) – an umbrella organization for prestigious national academies of science. This collaboration eventually ended as the IAP decided to produce *principles* that its member academies could incorporate into their own codes rather than the joint code envisioned by ICGEB. This was the case in large part because as an umbrella body of a diverse range of national academies, the IAP could not get all its member academies to adopt a code as such.

Result: The 2005 IAP *Statement on Biosecurity* provided five short principles to inform national academy codes.[17] By June 2005 ICGEB also

[16]United Nations. 2002. *Annex report of the policy working group on the United Nations and terrorism* A/57/273-S/2002/875, 6 August 2002. Available at http://www.un.dk/doc/A.57.0273_S.2002.875.pdf

[17]InterAcademy Panel. 2005. IAP statement on biosecurity, 7 Nov 2005 http://www.nationalacademies.org/morenews/includes/IAP_Biosecurity.pdf

decided to produce "building blocks" that others could draw on.[18] Neither set of advice included provisions setting out restrictions to access to knowledge and expertise as originally proposed.

A Universal Code?

Proposal: After the US rejection of a verification protocol for the Biological Weapons Convention in 2001, President Bush made a call for States Parties to the convention to "Devise a solid framework for bioscientists in the form of a code of ethical conduct that would have universal recognition."[19] At the US insistence, the BWC had as the topic for its 2005 meetings "the content, promulgation, and adoption of codes of conduct for scientists".[20]

Developments: By 2005, the US reversed its position to contend that a universal code would not be feasible.

Result: In 2005, possible considerations for a code were outlined in the final report of States Parties to the BWC.[21] In 2006, these states decided to re-examine this topic in 2008. As the BWC had a non-negotiating mandate for both the 2005 and 2008 meetings, no international code was agreed among states through the meetings and only several states reported on the national introduction of relevant codes.

I wish to draw out a number of observations from the initiatives in Box 8.1, points that characterize bioweapon codes discussions post-2001 more generally. One, this option has come up in various organizations as a way to guide the conduct of scientists. While professional codes with bioweapon-related components have been proposed in previous decades, the range of organizations involved and the extent of their attention to this option increased markedly after 9/11.[22] What has

---

[18]ICGEB. 2005. *Building blocks for a code of conduct for scientists, in relation to the safe and ethical use of biological science.* Trieste: ICGEB.

[19]Bush, G. 2001. *President's statement on biological weapons,* 1 Nov 2001, see http://www.whitehouse.gov/news/releases/2001/11/20011101.html

[20]Fifth Review Conference of the States Parties to the Convention on the Prohibition of the Development, Production and Stockpiling of Bacteriological (Biological) and Toxin Weapons and on Their Destruction. *Draft Decision of the Fifth Review Conference of the States Parties to the Convention on the Prohibition of the Development, Production and Stockpiling of Bacteriological (Biological) and Toxin Weapons and on their Destruction* BWC/CONF.V/CRP.3 6 November 2002.

[21]*Report of the Meeting of States Parties to the Convention on the Prohibition of the Development, Production and Stockpiling of Bacteriological (Biological) and Toxin Weapons and on their Destruction* BWC/MSP/2005/3 14 December.

[22]Rappert, B. 2004. *Towards a Life Sciences Code: Countering the Threats from Biological Weapons.* Bradford Briefing Papers (2nd series) 2004; No. 13 See http://www.brad.ac.uk/acad/sbtwc/briefing/bw-briefing.htm

been sought has not been so much a list of definite do's and don'ts, but rather means of engendering a "culture of responsibility".

Two, just who should be responsible for realizing a code has been a matter of some importance, not least because of the mix of science and security-orientated organizations with a stake in the issues. As suggested by the examples in Box 8.1, those associated with science have been looked to to undertake action for themselves.

Three, the fragmented and partial manner in which science is professionally structured curtails the potential to devise the sorts of codes that exist in other domains where entry to a profession is routinely subjected to licensing requirements.

Four, different threats in need of attention have been identified. While some activities have focused on fairly traditional matters such as physical access to pathogens, others (such as the Working Group of the United Nations and Terrorism) have gone beyond this to include publication practices. Determinations of what needs to be included have turned on contentious matters such as the extent of (terrorist) threats from biological weapons and feasibility of (often basic) research findings facilitating new capabilities.

8.4 What Has Been Accomplished?

For those looking for evidence of codes working as guides for conduct across the troubling waters where science and security intersect, experience in recent years might well be regarded as worrisome.

8.4.1 Codes As Exercises in Deferral

A possible source of concern would be the widespread practices evident of what might be called organizational "deferral". One way this manifested itself is in the secondary, advisory quality of many "codes". As alluded to in Box 8.1, prominent international players – such as the IAP, ICGEB, OECD, the BWC as well as the International Committee of the Red Cross and the International Union of Microbiological Societies – elected not to bring about a code but to provide advice to others about possible content for one. While this can be regarded as a prudent step to leave specific codes to those bodies closer to day-to-day research, it has had problems as well. Not least, it is not clear that many other organizations have acted on the advisory calls of these organizations. The only exception known to the author is the IAP *Statement on Biosecurity*. To date, among the nearly 100 member academies of the IAP, one follow-on code has been produced by the Royal Netherlands Academy of Arts and Sciences.[23] Technically, however, that biosecurity code is itself not a code as such, but again a set of provisions that research and business organizations

[23]Royal Netherlands Academy of Arts and Sciences. 2007. *A code of conduct for biosecurity.* Amsterdam: KNAW.

in that country could draw on for their own code – though again, it is not clear any have done so.

Deferral is evident in other respects. Consider the example of the National Advisory Board for Biosecurity (NSABB). It was set up in 2004 to provide advice to the US federal government about responses to concerns about how the findings and techniques of modern research might be misused. NSABB established a number of Working Groups to deliberate options and provide recommendations on a national institutional oversight framework, science communication, codes, synthetic biology, and international outreach. At the 13 July 2006 meeting of NSABB, *Considerations in Developing a Code of Conduct for Dual Use Research in the Life Sciences* was agreed by members of the Board.[24] As suggested by the title, this document does not provide a code or even recommend that one be adopted by relevant US agencies. Rather, as with the initiative mentioned above, it outlines considerations others could take up. As above, as far as is known to the author, no organization has acted on this to date.

Moreover, through the *Considerations* document, the NSABB Codes Working Group deferred responsibility to the other working groups. This was because the "shoulds" given almost wholly consisted of reiterations of the need for measures that were to be advised upon by the other Working Groups as specified in the Board's original 2004 charter. For instance, researchers were asked to assess the dual use potential of their research, which another working group in NSABB was devising advice about, and so on. Rather than adjudicating on any thorny matters, the provisions in *Considerations* restated the need to address them.

8.4.2 Follow Through?

A number of codes have sought to provide specific guidance on contentious issues. For instance, a group of NGOs lead by the Federation of American Scientists delineated what was permissible in biodefense programs.[25] Somerville and Atlas' nine-point *Code of Ethics for the Life Sciences* provides succinct ethical points intended to promote reflection about what constitutes responsible science.[26] Yet, none of these advocacy-orientated codes have been adopted more widely.

---

[24]http://www.biosecurityboard.gov/pdf/NSABB%20Draft%20Guidance%20Documents.pdf

[25]In November 2002 the Federation of American Scientists, Stockholm International Pease Research Institute, Verification Research, Training and Information Center, International Network of Engineers and Scientists for Global Responsibility, Acronym Institute for Disarmament Diplomacy, Sunshine Project, Pax Christi International, Physicians for Social Responsibility, and 20/20 Vision agreed draft recommendations for a code of conduct for biodefence programs. These were published as an Annex to Barbara Hatch Rosenberg, Defending against biodefence: the need for limit. *Disarmament Diplomacy*, Issue No. 69, February – March 2003. Available at: http://www.acronym.org.uk/dd/dd69/69op03.htm

[26]Somerville, M., and R. Atlas. 2005. Ethics: a weapon to counter bioterrorism. *Science* 307: 1881–1882.

The aforementioned paragraphs should not be taken to imply that no science organizations have adopted a bioweapons-related code. Yet, summary comments can be made about these initiatives that raise questions about their ultimate importance. One, such science codes have tended to consist of very short statements that provide little extension of current accepted standards, national regulations, or international laws. For example, in 2005 the International Union of Biochemistry and Molecular Biology agreed to a *Code of Ethics* which stipulated that members would not "engage knowingly in research that is intended for the production of agents of biological warfare or bioterrorism, nor promote such agents."[27] The 2005 *Code of Ethics* of the American Society for Microbiology underscored that bioterrorism was abhorrent.[28]

Two, where adopted codes have been more elaborated, it is not clear that many have been implemented with any conviction. So while the American Medical Association's *Guidelines to Prevent Malevolent Use of Biomedical Research* within its professional codes makes relatively detailed recommendations for new safeguards and oversight mechanisms for research,[29] their adoption and enforcement seems to have received little prioritization within the Association.

8.5 Reframings

To summarize the argument so far, against the said revolutionary advances taking place in the life sciences, codes have been forwarded as a central component of the policy responses. Numerous organizations associated with the governance of research have deliberated the whys and hows of bringing in a code since 2001. And yet, for all the activity that has taken place, little by way of concrete accomplishments relevant to practitioners can be identified. Judged on the basis of the number of codes, their effects in changing behavior, or their importance for the refinement of normative standards, progress to date would almost assuredly be found wanting. "High input for low output" would be one précis. As a result, the suggestion that codes have helped scientists kept pace with concerns about the potential of research to spread disease seems rather fanciful.[30]

Such a product oriented evaluation marginalizes the wider functions served by codes. Rather than looking at code documents and then asking whether they

---

[27] http://www.iubmb.unibe.ch/Standing_Orders/Code_ethics.htm

[28] http://www.asm.org/ASM/files/ccLibraryFiles/FILENAME/000000001596/ASMCodeofEthics05.pdf

[29] See Green, S., S. Taub, K. Morin, and D. Higginson. 2006. Guidelines to prevent malevolent use of biomedical research. *Cambridge Quarterly of Healthcare Ethics* 15: 432–439.

[30] For another instance of the reading of responsive measure as an effort in the 'simulation of control', in this case the risk-benefit analysis of the security implications of research, see Rappert, B. 2008. The benefits, risks, and threats of biotechnology. *Science and Public Policy* 35(1): 37–44, Feb.

have helped direct the behavior of practitioners, the process-oriented aspects of the deliberation about codes can be highlighted – what can be coined "codes talk".

This talk has served to enroll individuals and organizations into a certain (and for some rather novel) set of issues. For instance, the deliberation about codes within the IAP has acted to signal a level of unease within and outside of it regarding the potential destructive use of science. This, in turn, has provided an opportunity for those concerned within member academies to make a place for the biosecurity issues within crowded agendas. So, while preciously few codes have been produced, the author has collaborated with those in the science academies of Ukraine, United Kingdom, Israel, Uganda, and elsewhere, all of which have internally cited the IAP's *Statement on Biosecurity* to justify dedicating time and energy to this topic.

Further to these process benefits of "codes talk," as mentioned previously, the Royal Netherlands Academy of Arts and Sciences is the only national academy to date that has formally adopted a code on the back of the IAP's *Statement on Biosecurity* – in fact, a "code" better characterized as a set of considerations for others' codes. And while little has emerged in the way of follow-on outputs, the issue of devising a code has provided the focal topic for a series of consultations with practicing scientists and others held in the Netherlands by the Dutch academy.[31] Within these settings, awareness raising about the security dimensions of science has been a prime goal.

Much the same process aspects could be said about the 2005 and 2008 meetings of the BWC where science organizations with previously little engagement in international arms control participated in discussions about the security issues associated with life science research. In these settings, the topic of codes for *scientists* had the additional benefit of making the case for opening access to the BWC proceedings beyond those in government traditionally concerned with national security.

The previous claims suggest a certain placeholder function being fulfilled in recent years: the topic of codes provides a convenient one for bringing varied people together to discuss how to prevent the destructive applications of research. In this sort role, "codes" open a curious space. It is a space in which "everything" and "nothing" is at stake. With regard to "everything" – in raising questions about what constitutes proper conduct, codes talk provides an envelope for speaking to a wide range of issues and invites questions about how to set normative standards. Most people can find something to contribute regarding what should count as standards for conduct. Animated conversation can quickly turn into heated disagreement though as different ethical presumptions and pragmatic goals are traded.[32] To the extent code talk is treated as an occasion for trying to settle debate about what does and does not constitute acceptable conduct, then it is about a great many things.

---

[31] Van der Bruggen, K. 2009. Science of mass destruction. In *Biosecurity,* eds. B. Rappert and C. Gould. London: Palgrave.

[32] As in the debate codes in synthetic biology. See Check, Erika. 2006. Synthetic biologists try to calm fears. *Nature* 441, 388–389 and Etc. 2006 'Global Coalition Sounds the Alarm on Synthetic Biology' News Release 19th May. Available at http://www.etcgroup.org/en/issues/synthetic_biology.html.

With regard to "nothing" – given the modest accomplishments with codes in the past and the lack of significant prospects for them in relation to bioweapons, the talk is not likely to be consequential either. As a series of discussions rather than a movement towards binding proscriptions, codes talk has been rather limited too.

8.6 Evaluating the Process

Just what assessment should be given of the process-based dimensions of codes is a matter open to interpretation. On the negative side, the failure of the focus on codes to live up to the promise of guiding behavior and the lack of policeable standards agreed upon could be seen as quite problematic.[33] Against the suggestion that codes talk has served as a basis for achieving alternative aims – such as raising awareness – it could be countered that it would be more appropriate to undertake activities that directly set out to achieve these goals. The attention to codes in recent years could be said to be not a very efficient means to achieve notionally secondary ends or, more critically, a distraction eating up limited resources.

In contrast, the sympathetic reading could treat codes talk as part of an iterative effort to enroll more groups in attempts to counter threats from biological weapons. Given the relative absence of engagement by many science organizations into this matter in the past, any efforts that achieved significant traction could be judged as positive. In the case of the BWC, the active participation by groups traditionally outside of diplomatic arms control had the additional benefit of reinvigorating that convention. With the building up of community commitment achieved to date, it is possible to move further ahead in the future.

A major impediment to choosing between these negative and positive evaluations is uncertainty about what will happen in the future. What codes talk has enabled is still to be decided. Whether with the hindsight of history codes deliberations will be judged a sterile dead end or a stepping stone that (eventually) lead to highly significant activities cannot be resolved now; it depends on the twists and turns of what is to come.

Yet this situation presents an awkward trouble for those deciding about their participation in codes talk *now*: how is it possible to decide whether to press on with current efforts? In other words, when can it be said codes talk has degraded from a useful enrolling prelude that should be nurtured to instead become a stifling spinning of wheels that has gone on too long? Consider Box 8.2 in this regard. It characterizes and quotes some of the summarizing statements from controlled access meetings primarily dedicated to codes in the UK organized by the Foreign Office.

[33] Though at the time of writing, an industry association for the five leading German companies in the field of synthetic biology (Industry Association Synthetic Biology) issued a *draft* 'Code of Conduct'. See http://daccessdds.un.org/doc/UNDOC/GEN/G08/644/46/PDF/G0864446.pdf?OpenElement

Box 8.2 Meetings About Codes in the British Foreign Office

2003 – Initial seminar of those across government, academia and industry in the UK to discuss "utility, scope, promulgation, implementation, reactions, enforcement and next steps." A background note to that seminar indicated that "The starting premise of this note is that a code of conduct *is* a desirable objective because it can play a significant role in raising awareness of the [BWC's] prohibitions among the scientific and engineering community. . .A code of conduct is not a panacea, but one tool amongst many for combating BW proliferation. . . .The main issues here are what might a code contain and how might it be taken forward."[34]

2004 – Follow up seminar wherein a key message was that "further work on codes should build on existing systems, but that an overall statement of core principles could be developed as a guide for such continuing work."[35]

2005 – Meetings of the Biological and Toxin Weapons Convention that lead to advisory provisions by States Parties for organizations to take up as deemed appropriate.

2006 – Meeting in which it was decided "The imperative was to keep the issue alive and under discussion. It was encouraging to know that the general consensus was that, if embedded in existing systems and both feasible and proportionate, codes of conduct had a utility."[36] The summary of the session indicated that future work still to be undertaken would "ensure the appropriate progress continued."[37] This included that "there might be some role for Government in the production of suitable educational material but that the process of raising awareness and education in the science community should not be led by Government in the UK."[38] It also included that "the participants thought it would be useful to have a repetition of the seminar next year, possibly including international partners."[39]

---

[34]Emphasis in original. See FCO. n.d. *FCO Discussion Paper on a Possible Biological and Toxin Weapons Convention (BTWC) Code of Conduct.*

[35]FCO. 2004. *Biological and toxin weapons convention: Code of conduct. Lancaster house seminar 15 December 2003 Main Points.* London: FCO: 2.

[36]United Kingdom. 2006. *Codes of Conduct for Scientists* Sixth Review Conference of the States Parties to the Convention on the Prohibition of the Development, Production and Stockpiling of Bacteriological (Biological) and Toxin Weapons and on Their Destruction BWC/CONF.VI/WP.23 Geneva, 13–24 June 2005: 22 November: 2.

[37]Ibid., 2.

[38]Ibid., 2.

[39]Ibid., 2.

2008 – Meeting considering "lessons from history; current activities in academia and industry; government initiatives; international aspects; and lessons from [chemistry] on educational aspects that might be relevant for the BTWC."[40] It was concluded that "The UK recognises that codes of conduct for scientists and awareness raising campaigns do not offer a foolproof defence against the misuse of the life sciences for hostile purposes. But what they can do – along with measures on oversight and education – is to heighten the levels of awareness in the academic and research communities of the need for care; highlight the nature of the Convention's legal prohibitions; and promote the need to address issues such as technology governance on a continuing basis. Such issues cannot be dealt with quickly; sustained efforts by a broad range of stakeholders are required over an indefinite period of time."[41]

Three elaborations can be offered that extend the points raised in Box 8.2 and also relate the points to the broader themes of this chapter. One, in a certain respect the developments read in the reverse order of what would be expected if the process were advancing ahead toward some significant achievement. In 2003, the premise was that a code for (British) life science was a good idea that should be taken forward. The question was how. By 2008, the discussion focused on what might well be taken as groundwork lessons that could be drawn from elsewhere. Yet, despite this reversal in fortunes, the topic of codes continues to function as a focal point.

Two, not so apparent from the quotes themselves but key to enrollment dimensions of codes talk has been the changing audience participation in the meetings. Particularly the 2006 meeting brought the attendance of individuals new to the topic (i.e., practicing researchers). Even if the themes across years were notably similar and the prospects for achieving the systematic code first envisioned in 2003 were markedly receding over time, a broadening range of individuals were being brought into deliberations. As such, at any stage there was always a sense of a justification for continuing a conversation, even if it had become repetitive for some participants.

Third, what achievements could be realized have had an overtly "to-be-decided" flavor that could not be limited by what had come so far. Much of the prospects still lay in the future. While it seems rather unlikely to the author to believe any additional code achievements will take place in the UK beyond what has happened to date,[42] this cannot be completely ruled out either. Future world events are likely

---

[40]United Kingdom. 2008. *Oversight of ewmerging technologies: Examples of UK approaches to responsible development of science.* BWC/MSP/2008/MX/WP.11 12 Aug, available at http://daccessdds.un.org/doc/UNDOC/GEN/G08/626/54/PDF/G0862654.pdf?OpenElement

[41]Ibid., 6.

[42]As in the Health Protection Agency. 2005. *Principles of good scientific practice.* London: HPA, Aug.

to determine to what extent the codes talk in the UK to date has set the conditions for noteworthy accomplishments.

As an additional point complicating assessments of activities in the UK, the "mutually reinforcing" benefits of code activities have gained greater prominence over time. Codes are not seen as an achievement in themselves so much as part of wider set of developments being nurtured now and that will need to be sustained into the future. Not only does this contextualizing complicate making evaluations of what has been achieved with codes specifically, it also signals how attention to the process importance of "codes talk" is becoming explicitly recognized within deliberations. "Keeping the conversation going" is being portrayed as part of what needs to be done to keep attention on the potential for hostile use of the life sciences within a range of audiences.

8.7 A Disruption

The previous argument might be taken to affirm a prime contention that has been made by others studying codes of conduct: the process aspects of devising them are often central. Gotterbarn spoke in highly evocative terms to the importance of their consultative dimensions with the remark that *a code is nothing, coding is everything*.[43] Thus, one lesson that might be drawn is that what might be termed the "informal" dimensions of codes need to be central in any evaluation.

Such a message sensitizes us to a set of issues that might otherwise be passed over in evaluating how codes help to pace S&T developments. Yet for me as a policy analyst who has partaken in many deliberations about codes around the world, this conclusion and the argument of the previous section misses a major consideration: the rife doubt evident within international deliberations about the prospects for what can be achieved through codes – either by the codes themselves as instruments for guiding behavior or by codes talk acting as a springboard for follow up activities.

The summary given of the British Foreign Office deliberations gives just one indication of how the starting value attached to codes post-9/11 rather quickly faded. And yet, despite recognized limits, the attention to codes has continued all the same.

In my estimation there are few candidates that could be nominated as "believers" about the importance of codes to prevent the hostile use of the life sciences, even fewer if one moves away from prepared statements to engage people in private dialogue. While governments of varied stripes and science organization representatives presenting official position papers in fora such as the 2005 BWC meetings might attribute codes with much importance,[44] in my experience this has not been matched in policy meetings with a less public face. Indeed, with the lack of evangelicals, the author as someone studying this option has been extended numerous invitations to assume such a role. My repeated experience of being positioned as an advocate

[43]Gotterbarn, D. 1999. Not all codes are created equal. *Journal of Business Ethics* 22: 81–89.
[44]Rappert, Brian. 2009. *Experimental secrets*. Lanham, NY: University Press of America.

was, in fact, a prime inspiration for writing an autobiographical-style monograph, a book that examined my history of engagement with recent code activities in order to comment on the dilemmas that can arise for social researchers in trying to undertake "policy relevant" research.[45]

With this observation about the widespread disbelief evident, some follow-on clarifications are worth making. The earlier proposal to examine "codes talk" as a process of enrollment might well be taken by some readers of this chapter to imply a certain type of orientation has been evident in policy discussions to date. So, for instance, it could be taken to suggest that a commonplace sentiment echoed is that with just a little more sustained effort, significant achievements could be obtained. Or that the argument is often put that codes provide especially fruitful foci for international discussions. Yet, certainly as part of the interactions I have participated in, such promissory and exceptionary claims are rarely made.

After the several years in which bioweapon-related codes have been debated, for me the matter in need of consideration is not just why potential is still invested in them despite the limited achievements to date, but also why have they been discussed this long despite the early and frequently expressed personal doubt expressed by individuals about their likely potential.

8.8 A Reconsideration

How can this situation be made sense of and what does it tell us about the policy process? Some might seek to explain the continuing attention despite the limited hopes for progress. This could be related to institutional inertia, geopolitics, the play of personalities, or a host of other factors.[46]

In the remainder of this chapter, instead of pursuing such lines of explanation, I want to use the consideration of codes as an occasion for reflecting on wider issues about the place of doubt within policy-making processes. These comments are speculative in nature, but they are based on my experience of partaking in numerous code-related initiatives over several years and across various national contexts.

The inspiration for the comments that follow comes from an analysis by the anthropologist Michael Taussig of the role of disbelief within shamanism.[47] As with many others studying religious faith, his starting point is seeking to understand how skepticism and belief are intertwined. Across various contexts, he suggests how shamans often foster doubt in their own power. Ways of healing are continually exposed as fraudulent. Yet, in practice, this did not have the effect of weakening

---

[45] Ibid.

[46] For an example of an analysis that attempts to determine why so much attention has been given to codes given their shortcomings, see Lentzos, Filippa. 2006. Managing biorisks: Considering codes of conduct. *Nonproliferation Review* 13(2) July: 221.

[47] Taussig, M. 2003. Viscerality, Faith, and Skepticism. Another Theory of Magic. In *Magic and Modernity,* eds. B. Meyer and P. Pels. Stanford, CA: Stanford University Press: 272–306.

belief in shamanism overall. Rather, the exposure of practices as "mere tricks" is part of a cyclic process. The unmasking of certain shamanistic practices as fakes encourages a search for the real secrets which are later exposed as fakery which then sets off a further search for authentic secrets, etc. With this complex treatment of faith, those taking part in shamanistic practices have a similarly complicated status. As Taussig contends, shamanism:

> relies on corrosive scepticism [. . .] in which scepticism and belief actively cannibalize one another so that continuous injections of recruits [. . .] are required. They are required, so it would seem, to test and therewith brace the mix by serving not as raw material of doubt positioned to terminate as believers, nor yet as cynical manipulators, but as exposer vehicles for confession for the next revelation of the secret contained in the trick that is both art and technique and thus real and really made up.[48]

The "judicious and intricately moving medleys of scepticism and faith" act to continuously defer the resolution of the ultimate power of shamanism.[49]

In suggesting how doubt can co-exist with belief (and even how it might be necessary for belief), Taussig provides a provocative way of understanding what is taking place with bioweapons-related codes. As with shamanism, codes can be approached as a process of enrollment, but critically not one where those participating should be seen as on course to becoming either believers or cynics. Rather the mix of belief and non-belief within individuals and collectively is part of what propels the process on.

So in my experience the expression of doubt about what can be accomplished through codes is almost always part and parcel of "codes talk". Yet voiced individual and collective apprehension about the limitations of codes and codes talk often results in calls for "more" – more people as part of the process, more wide ranging discussions, more varied codes, more considered action. It is difficult for me to believe many of those actively making such recommendations in national science academies, the BWC, or elsewhere can be understood through the labels of converts or cynics. Rather, as with the Foreign Office meetings surveyed through Box 8.2, they seem much more aptly described as engaged in a process mixing doubt and belief in a manner that sets the basis for future rounds of doubt and belief.

The way in which the topic of codes raises basic questions about the usefulness of ethical principles lends itself to the mixing of doubt and belief. Some ethicists and social scientists might worry about whether the general provisions typical in codes have been or ever could be useful for concrete action. Yet when skepticism and belief are seen as integral to maintaining a conversation, then the ready disagreement about the possible utility of codes can provide the base elements for carrying on with further discussion. Similarly, the magnitude of the challenges associated with preventing the destructive use of life sciences into the future offers much scope for debate about what (if anything) can be realistically done.

Another line of Taussig's argument is worth mentioning. He offers an interesting warning to those wishing to debunk shamanism as mere trickery: such acts help

---

[48]Ibid., 288.

[49]Ibid., 294

perpetuate belief. So, he recounts (with some humor) how anthropologists bent on exposing certain acts of healing as trickery have ended up fulfilling the unmasking role required for belief to continue. Likewise, I think it is useful to ask whether expressions of skepticism about codes are in practice highly functional in keeping the option alive as a matter for discussion. Since 2006, I have offered various critical publications, presentations and workshop interventions related to bioweapon codes. Yet, rather than reducing the standing of codes, I often wonder whether it has been having an opposite effect. At a basic level, such interventions help continue attention to this option. Because much of what "codes" are relates to the process aspects of their deliberation, this attention is quite significant in keeping codes alive as an option. That the interventions are critical in tone overall (without pretending to foreclose what will happen in the future[50]) is in many ways not that important because voiced skepticism is already widespread.

The eventual effects of any intervention are not something easy to gauge. With his focus on the perpetuation of cycles of belief and non-belief, Taussig does not address how the standing of shamanism has or could rise and fall over time. Faith appears as indestructible when it is constituted by conviction and doubt. Yet, patterns of belief do change over time. Presumably, if the future were to witness more and more criticism about codes and less and less hope, then they would come to be no longer regarded as options for consideration by so many. As part of asking how codes help pace science and technology, the purpose of this section has not been to settle the ultimate standing and direction of code discussions. Rather, it has been to suggest another, somewhat counter-intuitive, way of thinking about how continuing attention can remain with certain policy options despite widespread voiced skepticism.

8.9 Conclusions

Against the wide ranging concerns about the accelerating pace of S&T developments, the chapters of the volume have sought to identify perspectives and tools commensurate with the challenges faced. Across a range of topics, what are varyingly defined as "codes of conduct" have been identified by many governments and organizations as one such instrument. By way of considering the possible place of codes, this chapter has detailed the recent turn in attention to codes intended to prevent the destructive use of life science research.

In doing so a central aim has been to shift the discussion away from conventional standards for evaluating the way in which codes matter for meeting challenges. That has entailed moving from conceiving of codes as documents for policing the behavior of practitioners to codes as elements of processes for raising the profile of issues within and across organizations. Herein it might be said that what matters about

[50]While recognizing that what past and current codes discussion will enable in the future cannot yet be known, I have sought to raise doubts about expectations given past experience.

codes initiatives is the extent the activities fostered entail more than just devising written codes. Attending to these process-orientated aspects of codes for an emerging area of professional concern raises many thorny issues for evaluating what has been achieved and what sort of continuing effort should be invested in the future.

Moving beyond noting such difficulties though, this chapter has sought to acknowledge and make sense of the doubt about the potential of codes that has accompanied their (re-) emergence since 2001. As argued, rather than being perplexed about how the attention to codes could have continued as long as it has despite the many reservations associated with them, another way to conceive of the policy process is as imbued with the play of doubt and belief. How these mix together in practice is part of how the deliberation about codes has continued to date.

Chapter 9
An International Framework Agreement on Scientific and Technological Innovation and Regulation

Kenneth W. Abbott

9.1 Introduction

In this chapter, I consider *international* action to help law and policy better keep pace with emerging scientific and technological innovations. I focus especially on addressing innovations that have the potential to create significant health, safety and environmental risks and/or to pose significant social, cultural and ethical challenges, even as they promise significant benefits. Prominent among these are Brad Allenby's "five horsemen of emerging technologies"[1] – nanotechnology, biotechnology, robotics, information and communication technology, and applied cognitive science, along with their increasingly powerful combinations – as well as synthetic biology. Even more challenging, of course, are the equally disruptive innovations that have not yet been introduced, or of which we are not yet aware.

In speaking of "international action," I do not mean to suggest that innovations like those just mentioned can or should be directly regulated, or otherwise addressed by law, at the international level. Direct international regulation is not a realistic option, in large part because the international legal system is relatively weak in all three of the institutional forms Gary Marchant identifies as collectively constituting "law:" "legislation, regulation, and judicial case law."[2] First, there is no global legislature capable of adopting statutory law. Second, there is no global Food and Drug Administration, Environmental Protection Agency or similar administrative organ capable of adopting effective regulatory law. To be sure, rough analogues of regulatory agencies exist among the "specialized agencies" of the United Nations system – e.g., the World Health Organization (WHO) – and other international organizations, such as the UN Environment Programme (UNEP). A few of those entities, including the WHO, have even been granted authority to adopt "regulations" applicable to states (that is, to nations), albeit in narrow circumstances and typically subject to

[1]Brad R. Allenby, Chapter 1, this volume, at [8].

[2]Gary E. Marchant, Chapter 2, this volume, at [22].

K.W. Abbott (✉)
Arizona State University, Tempe, AZ, USA
e-mail: ken.abbott@asu.edu

G.E. Marchant et al. (eds.), *The Growing Gap Between Emerging Technologies and Legal-Ethical Oversight*, The International Library of Ethics, Law and Technology 7,
DOI 10.1007/978-94-007-1356-7_9, © Springer Science+Business Media B.V. 2011

opt-out rights.[3] Yet on the whole international organizations are so weak that they are hardly comparable to powerful domestic regulatory agencies. Few if any of them, moreover, have authority to adopt binding regulations applicable to private actors, including business firms and researchers. Third, in spite of a recent wave of judicialization,[4] the international system still relies on courts far less than do domestic legal systems, especially in common law countries. What is more, tribunals such as the International Court of Justice have traditionally not observed principles of *stare decisis* that enable them to develop binding "judicial case law."[5]

Given the limits on international legislation, administrative regulation and case law, we are left with the two traditional mechanisms of international law-making, customary law and treaties. Neither, unfortunately, is generally associated with the rapid action needed to deal with exponential rates of innovation.[6] Customary international law is developed through the actual practice of states, as well as the subjective acceptance by states of regularities in practice as legally binding rules ("*opinio juris*"). Customary law has traditionally been based on gradual accretions of practice, and has thus been quite slow to develop, although there have been many recent efforts to expedite the process.[7] Custom may also work best in areas that directly engage the state and respond to relatively simple rules, such as the breadth of the territorial sea or the immunity of diplomats. Formal treaty negotiations are also relatively slow and costly – although recent innovations in areas such as international environmental law have made the process considerably more flexible, as discussed further below. In addition, the treaty process has traditionally privileged diplomats and high executive officials over technical experts. In short, at least in their traditional forms, both mechanisms are even less well-suited than national law to the dynamic context of scientific innovation.

More precisely, then, the focus of this chapter is on international action to *coordinate national* law and policy in responding to scientific and technological innovations.[8] Beyond merely coordinating, moreover, international arrangements can *steer* national actions in desirable directions. "Steering" (in contrast to "regulation" or

---

[3]See, e.g., Constitution of the World Health Organization, Art. 21–22.

[4]See Goldstein, Judith, Miles Kahler, Robert O. Keohane, and Anne-Marie Slaughter. 2001. Introduction: Legalization and world politics. In *Legalization and world politics*, eds. Judith Goldstein, Miles Kahler, Robert O. Keohane, and Anne-Marie Slaughter. Cambridge, MA: MIT Press.

[5]See, e.g., Statute of the International Court of Justice, Art. 59: "The decision of the Court has no binding force except between the parties and in respect of that particular case."

[6]See Marchant, supra, at [19].

[7]Especially in rapidly developing areas such as human rights, advocates have sought to develop a more dynamic "modern" form of customary law by deemphasizing the accretion of state practice and instead emphasizing declarations of rules by states (which arguably reflect *opinio juris*) in fora such as the General Assembly, or even in multilateral treaties. See Roberts, Anthea Elizabeth. 2001. Traditional and modern approaches to customary international law: A reconciliation. *American Journal International Law* 95: 757.

[8]The introductory chapter by Gary Marchant, supra, implicitly assumes that legal responses to innovation will be national, and indeed focuses primarily on the US.

"coercion") refers to the use of persuasion, incentives, peer pressure and other "managerial" forms of influence[9]; here it would be designed to move national actions not only toward greater uniformity, but also toward greater efficiency, effectiveness, legitimacy and public-interest orientation in form and content.[10] Finally, international arrangements might *facilitate* speedier and better designed national legal and policy responses, especially in states with limited regulatory capacity.

I do not attempt to specify here the substantive content of international coordination, steering and facilitation – exactly which legal and policy responses international arrangements should encourage and support – or even the details of international organizations and procedures. Both inquiries are essential, but are parts of a larger project. Instead, I suggest here a general institutional framework for international coordination, steering and facilitation, consisting of two main elements. The first is an international "framework convention," a particularly flexible type of treaty prominent in international environmental law and certain other fields. While negotiating a framework convention involves some of the same problems of delay and cost as any treaty process, the framework convention model is explicitly designed to be initially less demanding and easier to negotiate, while facilitating more detailed incremental and adaptive rule-making over time, as better information on risks and benefits is obtained. The second element is a set of international institutions and procedures to coordinate, steer and facilitate national action, which would be established by and operate under the authority of the framework convention.

Even under existing framework conventions, rule-making and implementation often engage actors beyond the participating states as such. The arrangements I propose would build on this experience by incorporating as essential parts of the regime actors operating at three levels of governance.[11] The first is the traditional "international" or inter-state level. Here the framework convention and its basic institutions and mechanisms would be created and managed by representatives of the "state parties." The second is the "trans-governmental" level, which includes cross-national arrangements among executive agencies and officials, legislative bodies and legislators, and other units and individuals within national governments, rather than among "states" as such.[12] Here the convention would authorize and rely for rule-making and implementation on cooperation among national regulatory agencies, regulators and other relevant units and officials within "the state." The third is the "transnational" level, which includes cross-national relationships among societal actors and organizations. Here the convention would empower and rely on cooperation among

---

[9]On the managerial approach, see Chayes, Abram, and Antonia Handler Chayes. 1998. *The new sovereignty: Compliance with international regulatory agreements.* Washington: Brookings.

[10]See, e.g., Wood, Stepan. 2002–2003. Environmental management systems and public authority in Canada: Rethinking environmental governance. *Buffalo Environmental Law Journal* 10: 129.

[11]For a useful introduction to the three levels of governance identified here, see Pollack, Mark A., and Gregory C. Shaffer. 2001. Transatlantic governance in historical and theoretical perspective. In *Transatlantic governance in the global economy*, eds. Mark A. Pollack and Gregory C. Shaffer. Lanham, MD: Rowman & Littlefield Publishers. Inc.

[12]Slaughter, Anne-Marie. 2004. *A new world order.* Princeton, NJ: Princeton University Press.

private actors that have stakes in the issues of concern and can contribute to effective legal and policy responses. Two essential societal groups would be actors engaged in scientific and technological R&D and science- and technology-based business firms: these actors produce the innovations to which law must respond, are most knowledgeable about those innovations, and have the authority, access and information to produce meaningful compliance with legal rules on a day-to-day basis in the lab or factory. Other important societal actors are those potentially affected by innovations – e.g., workers and consumers – and representatives of the larger (transnational) society concerned with the social, cultural and ethical implications of innovations.

9.2 Benefits of International Coordination

Given the difficulties faced by even a highly-developed national legal system like the US in keeping pace with exponentially accelerating innovation, why should we consider introducing the additional complexities of international action? The value of international coordination, steering and facilitation arises out of two sources: the underlying problem of keeping pace with innovation, and the secondary problem of inconsistent or inappropriate national responses.

In terms of the underlying problem, keeping pace with emerging scientific and technological innovations depends crucially on information: legal and policy institutions require early warning of significant innovations; information on new technologies and their potential benefits, risks and other impacts[13]; means of assessing unclear data; even some degree of prediction.[14] In all these areas, more information is better than less, and multiple social, cultural and ethical perspectives are more valuable than unitary, possibly myopic ones. International coordination, steering and facilitation can help states and societal actors produce and share information more effectively, while increasing the comparability of information and assessments from varied sources, e.g., in terms of nomenclature, metrology, indicators and presentation formats. International action can also enhance the diffusion of information from the first movers in an area of innovation to those at earlier stages of the technology life cycle. In addition, scientific and technological innovations can have significant cross-border effects: on the environment, human health or even national security in other states. International coordination helps guarantee a flow of information to affected states and societal actors – both about the effects themselves and about the nature and results of any control measures taken by the first-movers – improving their ability to respond in a timely and effective way.

[13]Currently, as Allenby notes, "the public (and, indeed, technologists outside of their particular specialties) get only impressionistic glimpses of emerging technologies in stories about particular events or experimental results." Allenby, supra, at [9].

[14]Predicting the full range of impacts of a significant scientific innovation at an early stage is, of course, impossible. See Allenby, supra, at [7].

In terms of national responses, inconsistent national regulations are frequently a drag on economic activity, especially on international trade. The standard example is the conflict between the US and Europe over genetically modified foods (GMOs). From the US perspective, relatively strict European regulation functioned as a trade barrier. The US and other agricultural exporting countries challenged certain restrictions on GMOs by the EU and individual EU member states through the dispute settlement process of the World Trade Organization (WTO); in 2008 the US went so far as to request WTO authorization for trade sanctions in response to the EU's failure to implement the resulting decision.[15] Economic conflicts like this can harm international relations more generally, interfering with welfare-enhancing cooperation in other areas. Inconsistent national regulation also hampers scientists, engineers and business firms who seek to collaborate across borders on research, development and commercialization.

Beyond merely reducing inconsistency, international action that incorporates an element of steering can help offset incentives that lead states to over- or under-regulate. Overall, the stronger incentives are probably to *under*-regulate, largely for competitive reasons. In highly dynamic fields such as the "five horsemen," researchers and firms compete strenuously to discover, develop and commercialize scientific and technological innovations; governments compete almost as strenuously to support, finance and eliminate regulatory obstacles for national innovators: for the direct economic gains, the positive externalities innovation can create for other economic sectors, and national prestige. In terms of regulation, this is a classic Prisoner's Dilemma, akin to the well-known but controversial "race to the bottom" in areas such as worker rights and environmental protection.[16] The incentives to under-regulate are enhanced when a technology creates negative trans-border externalities, passing some of the costs to others. In these situations, international steering can play a significant role in restraining national "defection."

Apart from the stringency of regulation, it is widely recognized that states need encouragement (steering) and support (facilitation) to move toward "better regulation." For example, the European Commission has adopted a comprehensive "better regulation" strategy to improve regulatory actions at EU level and within its member states.[17] The aim of such efforts is to encourage regulatory approaches and instruments that are more efficient and flexible, less costly, more effective, and

---

[15] Arbitration between the parties on the appropriateness of US retaliation was suspended to give the EU a longer time to implement the decision. For a summary of the dispute settlement proceeding – European Communities – Measures Affecting the Approval and Marketing of Biotech Products, WTO Dispute DS291 – see http://www.wto.org/english/tratop_e/dispu_e/cases_e/ds291_e.htm, visited 10 Sept 2009.

[16] See, e.g., Prakash, Aseem, and Matthew Potoski. 2006. The voluntary environmentalists: Green clubs, ISO 14001 and Voluntary Regulations 352–53 and sources cited (2006).

[17] See http://ec.europa.eu/governance/better_regulation/index_en.htm, visited 10 Sept 2009. EU member states, such as the United Kingdom, have empowered lead agencies to promote the better regulation agenda across the national government. See http://www.berr.gov.uk/whatwedo/bre/, visited 10 Sept 2009.

more appropriate in terms of their social, cultural and ethical impacts than traditional forms of legal action. Better regulation in one state not only helps that state deal effectively with local innovation risks, but also leads it to moderate any adverse trans-border effects. To the extent that better regulation programs lead states to rely on comparable regulatory techniques, moreover, they enhance international coordination.

Finally, many economies still classed as "developing" or "in transition" engage in sophisticated scientific and technological research on a large scale. The obvious example is China, "now a key global player in R&D in terms of absolute size as well as growth rates."[18] Yet many such states, and even some more advanced countries, lack the regulatory capacity to foresee and respond adequately to emerging risks. International institutions can facilitate effective and socially appropriate responses in these states (while reducing potentially harmful trans-border effects and incentives to under-regulate) by providing low-cost information, disseminating effective regulatory models, and granting other forms of technical and financial assistance. At the same time, such support can facilitate home-grown innovation within developing and transitional economies as well as the dissemination of appropriate innovations to them, helping to address the narrowing but still substantial North-South divide in science and technology.[19]

To be sure, states differ widely in their economic interests, "risk cultures,"[20] social relations, legal and regulatory systems, levels of development, capacities and other attributes. As a result, it is inevitable and often appropriate that states will respond to particular innovations with different forms and levels of regulation. For example, where the principal risk of an innovation is the disruption of traditional practices or ethical principles that vary across cultures, an international regime might appropriately stop at encouraging the exchange of information, assessments and cultural viewpoints; stronger efforts to promote uniform regulation would be counterproductive.[21] Again, then, the goals of an international arrangement must be coordination, steering and facilitation – even true "harmonization," let alone direct international regulation, is very unlikely and might be inappropriate.

Even in these limited terms, current international arrangements are far from adequate; indeed, their weaknesses are probably more severe than any of the domestic

[18]See OECD. 2007. Innovation and growth: Rationale for an innovation strategy 7–8, available at http://www.oecd.org/dataoecd/2/31/39374789.pdf, visited 10 Sept 2009.

[19]See World Bank, Global Economic Prospects 2008: Technology Diffusion in the Developing World (2008).

[20]See International Risk Governance Council. 2008. An introduction to the IRGC risk governance framework 7, available at http://www.irgc.org/The-IRGC-risk-governance-framework,82.html, visited 10 Sept 2009.

[21]Some states might still have an incentive to under-regulate, e.g., to attract individuals seeking to access a medical technology that is restricted in their home states. Yet inconsistent national regulation could also be an effective sorting mechanism in these situations, allowing nations to regulate their own societies in accordance with their dominant cultural beliefs, while still allowing individuals that do not share those beliefs to access the regulated technologies. Thanks to Lyria Bennett Moses for suggesting these points.

problems discussed in other chapters of this volume. Most states and national regulatory agencies do not coordinate their responses to risky innovations at all. Many of the collaborative arrangements that do exist focus primarily on promoting innovation, not on analyzing or responding to risks or to social or ethical concerns; an important example is the effort by the OECD Directorate for Science, Technology and Industry to develop a strategy for promoting innovation in member countries.[22] Steering is even more limited, with important exceptions such as the European Union better regulation agenda.[23] Most current arrangements emphasize streamlining or reducing regulation[24]; when regulation is discussed, moreover, the focus is on current issues with immediate economic impact, not on emerging or future technologies and risks. Relatively modest efforts could significantly improve the international science and technology regime.

9.3 Uncertainty

The core feature of the "pacing law with science and technology" problem, whether at the national or international level, is its inescapable uncertainty: "Uncertainty is pervasive in risk regulation, by definition."[25] Fundamentally, regulatory authorities must try to *anticipate future developments*, including the potential benefits, risks and other impacts of early-stage innovations. More specifically, at least three types of uncertainty frequently exist.

First, most innovations pose various forms of *technical uncertainty,* both static and dynamic, relating to the ultimate nature and scope of the innovation itself and its potential benefits, risks and impacts.[26] Second, many of the innovations discussed in this volume involve *normative uncertainty*: at least some individuals find it difficult to reconcile what they know of a technology with personal values or prevailing social norms; they may even find it difficult to determine which values and norms should apply.[27] Technologies for cognitive enhancement, radical lifespan

---

[22]Early work on this strategy suggests minimal consideration of regulatory issues. See http://www.oecd.org/pages/0,3417,en_41462537_41454856_1_1_1_1_1,00.html, visited 10 Sept 2009.

[23]See note 17, supra.

[24]See, e.g., the OECD. 2005. Guiding principles for regulatory quality and performance, available at http://www.oecd.org/document/38/0,3343,en_2649_34141_2753254_1_1_1_1,00.html, visited 10 Sept 2009. A major tool of streamlining is regulatory impact analysis (RIA), which enables cost-benefit analysis of proposed regulations. See http://www.oecd.org/document/49/0,3343,en_2649_34141_35258801_1_1_1_1,00.html, visited 10 Sept 2009.

[25]See OECD. 2006. Working party on regulatory management and reform, risk and regulation: Issues for discussion 7, available at http://www.oecd.org/dataoecd/20/39/37551219.pdf, visited 10 Sept 2009.

[26]For example, the International Risk Governance Council, supra note 20 at 16–17, distinguishes *complex* problems (where the cause of particular observed effects is uncertain because many causal agents operate simultaneously) from *uncertain* problems (where the data on benefits or risks are insufficient or unclear).

[27]Values and norms also differ widely between communities, even within a particular state.

extension and synthetic biology all create challenging problems of normative uncertainty, contributing to their "wicked complexity."[28] Third, technical and normative problems create *political uncertainty* for legal and policy officials. At early intervention points in the life cycle of an innovation, policy-makers cannot be sure what regulatory measures will be effective against the likely risks, and what the costs and effects of alternative measure may be. Even more important, policy-makers cannot easily predict which if any interest groups or "value actors"[29] will oppose the innovation – as some activists oppose GMOs and nanotechnology[30] – or contest particular government responses, or how the general public will react. Without an understanding of the potential political costs and benefits, few policy-makers will willingly implement a meaningful regulatory response.

Given the centrality of information problems, an international arrangement must include certain essential elements:

(a) It should steer states toward more extensive production, sharing and assessment of information about potential benefits, risks and impacts, and toward the adoption of comparable nomenclatures, metrology, indicators and methodologies.
(b) It should promote technical, social and political learning and normative deliberation to overcome the three forms of uncertainty, both within and across states.
(c) It should promote and facilitate the adoption of comparable, effective and efficient national regulatory frameworks, including both general science and technology policy and specific regulatory mechanisms.
(d) It should be capable of relatively rapid action as problems appear, and flexible enough to recalibrate those actions as new information and understandings emerge.
(e) Finally, it should encourage the participation and engagement of stakeholders, relevant epistemic communities and civil society, both within states and transnationally: to facilitate learning and deliberation, minimize political disputes, avoid regulatory capture, and further democratic principles.

These are challenging requirements given the difficulties of international legal action. The framework convention model is expressly designed to meet them at

[28]See Allenby, supra, at [12]. The IRGC, supra note [20, at 16–17], refers to these as *ambiguous* problems, which involve "divergent or contested perspectives on the justification, severity or wider meanings associated with a given threat."

[29]See Abbott, Kenneth W., and Duncan Snidal. 2002. Values and interests: International legalization in the fight against corruption. *Journal of Legal Studies* 31 (1): Part 2, S141–78.

[30]For example, the ETC Group proposes a "moratorium ... on [nanotechnology] research involving molecular self-assembly and self-replication" as well as strong public oversight of all nanotechnology development; it recently sponsored a design contest for a universal "nanohazard" symbol. http://www.etcgroup.org/en/issues/nanotechnology.html, visited 10 Sept 2009. As to GMOs, the ETC Group argues that "in the current social, economic and political context, genetic engineering is not safe, and involves unacceptable levels of risk to people and the environment." http://www.etcgroup.org/en/issues/biotechnology.html, visited 10 Sept 2009.

the inter-state level. To fully meet these requirements, however, an international arrangement must incorporate "governance as well as government:"[31] bringing into the regime the capabilities of public and private officials, experts and stakeholders; and adopting a broad understanding of "law" that includes public and private "soft law," such as codes of conduct, guidelines and best practices.[32]

9.4 A Framework Agreement on Scientific and Technological Innovation and Regulation

An international system to coordinate, steer and facilitate national responses to emerging innovations requires a "constitution." To ensure the stability of the system, this should take the form of a treaty, the basic legal structure of international regimes. If treaty negotiations proved too costly or provoked strong resistance, states could set out the fundamental principles and procedures of the new system in a non-legally binding instrument, such as a declaration of an international organization; however, the treaty form offers significant advantages, discussed further below. The treaty should be structured as a framework convention to ease initial negotiations and provide for adaptive rule-making over time. We might call this the Framework Agreement on Scientific and Technological Innovation and Regulation (FASTIR) – a title whose acronym evokes the desire for speedy legal responses.

9.4.1 Innovation

As its title suggests, FASTIR could include arrangements to facilitate scientific and technological innovation within participating states. A possible basis for such arrangements might be the work of the Organization for Economic Cooperation

---

[31]Cf. Rosenau, James N., and Ernst Otto Czempiel, eds. 1992. *Governance without government: Order and change in world politics*. Cambridge: Cambridge University Press. For a recent application of governance theory to the regulation of business firms in areas such as worker rights, human rights and the environment, see Abbott, Kenneth W., and Duncan Snidal. 2009. Strengthening International Regulation through Transnational New Governance: Overcoming the Orchestration Deficit. *Vanderbilt Journal of Transnational* 42: 501. See also Mandel, Gregory N. 2008. Nanotechnology governance. *Alabama Law Review* 59: 1323.

[32]On public soft law, see Abbott, Kenneth W., and Duncan Snidal. 2000. Hard and soft law in international governance. *International Organization* 54(3): 421. On private soft law, see Abbott & Snidal, Strengthening International Regulation, supra; Cashore, Benjamin, Graeme Auld, and Deanna Newsome. 2004. *Governing through markets: Forest certification and the emergence of non-state authority* (referring to norms of influential private institutions as "private sector hard law"); Kirton, John J., and Michael J. Trebilcock. 2004. Introduction: Hard choices and soft law in sustainable global governance. In *Hard choices, soft law: Voluntary standards in global trade, environment and social governance,* eds. John J. Kirton and Michael J. Trebilcock (defining soft law as relying "primarily on the participation and resources of nongovernmental actors in the construction, operation, and implementation of a governance arrangement").

and Development (OECD). As mentioned above, the OECD has since 2007 been developing a general "innovation strategy" for its member states – virtually all of which are advanced industrial economies – as a way to promote sustained economic growth.[33] The OECD is due to complete its strategy in 2010; FASTIR could incorporate its major principles and establish mechanisms to promote them. FASTIR could also promote cross-border collaboration in research and development, another area in which OECD has done useful work.[34] Finally, FASTIR could incorporate mechanisms to promote indigenous innovation within developing countries, as well as the diffusion to such countries of appropriate innovations created in the North.[35]

9.4.2 The Framework Convention-Protocol Approach

In this chapter, however, I focus not on encouraging innovation, but on mechanisms for responding to the potential risks and adverse impacts of emerging innovations through "regulation," broadly defined.[36] As a "framework convention" FASTIR would have three principal functions relevant to regulation:

- First, it would establish the "pacing law with science and technology" problem as a legitimate issue of international concern,[37] while providing structures for ongoing international cooperation. As a "framework" convention, FASTIR would not itself address specific innovations or resolve specific issues; it would thus include few substantive obligations. It should, however, set forth agreed *objectives, principles* and *general commitments* to guide national and collective action over time, and should establish *institutions* and *procedures* to coordinate, steer and facilitate national legal and policy actions.

---

[33]See http://www.oecd.org/pages/0,3417,en_41462537_41454856_1_1_1_1_1,00.html, visited 10 Sept 2009. See also Gault, Fred, and Susanne Huttner. A Cat's Cradle for Policy. *Nature* 455: 462, 25 Sept 2008.

[34]See http://www.oecd.org/document/10/0,3343,en_2649_34319_35044426_1_1_1_1,00.html, visited 10 Sept 2009.

[35]Achieving these goals might require somewhat different approaches than those used in advanced economies, although certain approaches would be fruitful in both settings. See Sarah Box. 2009. OECD Work on innovation – A stocktaking of existing work. DSTI/DOC 2: 46–48, available at http://www.oecd.org/dataoecd/14/32/42095821.pdf, visited 10 Sept 2009. See generally World Bank, Global Economic Prospects 2008, supra note 19.

[36]There have been many efforts to define "regulation" to encompass trans-governmental and transnational standards and procedures as well as traditional mandatory state regulation. For example, Julia Black, Enrolling Actors in Regulatory Systems: Examples from UK Financial Services Regulation, 2003 Pub. L. (Spring) 63, 65 (2003), defines "regulation" as "the sustained and focused attempt to alter the behavior of others according to defined standards or purposes with the intention of producing a broadly identified outcome or outcomes, and which may involve mechanisms of standard-setting, information-gathering and behavior-modification."

[37]Daniel Bodansky. 1999. *The framework convention/protocol approach framework convention.* Tobacco control technical briefing series, No. 1, WHO/NCD/TFI/99.1, p. 20, available at http://www.who.int/tobacco/resources/publications/fctc/en/, visited 10 Sept 2009.

- Second, it would establish procedures by which states could create detailed substantive agreements ("protocols," "annexes" or other subsidiary instruments) when conditions became ripe for concrete international action. Such instruments could address specific fields of innovation (e.g., nanotechnology or synthetic biology) as well as cross-cutting issues (e.g., public input on ethical controversies). FASTIR would thus initiate – or at least facilitate – an incremental process of rule-making, enabling states and other actors to take more definitive, coordinated action as they gain sufficient information and understanding to overcome technical, normative and political uncertainty. This incremental process is known as the framework convention-protocol approach.
- Third, FASTIR would establish mechanisms to encourage and facilitate the acquisition and sharing of information on emerging innovations, as well as forms of dialogue and deliberation designed to produce deeper understanding of the risks, benefits and social and ethical implications of those innovations.

To properly perform these three functions, FASTIR should be a legally binding treaty. Even though it would lack concrete substantive obligations, its binding legal character would commit states to common principles, institutions and procedures; obligate them to develop and share information and assessments and participate in dialogue; and establish a firm basis for subsequent coordinated rule-making.[38]

9.4.3 General Provisions

Two prominent framework conventions – the 1988 Vienna Convention for the Protection of the Ozone Layer and the 1992 United Nations Framework Convention on Climate Change (FCCC) – suggest certain general provisions that FASTIR should include. First, the agreement should enunciate basic *principles to guide future action.* Such principles would inform the ongoing dialogue among participating states and other actors, reduce the transactions costs of later negotiations, and help shape national actions – both directly, as states accept and internalize the principles, and indirectly, as other states and domestic interest groups pressure states to observe them.

Based on Vienna and the FCCC, participating states might commit to:

i. cooperate in good faith to facilitate and coordinate the acquisition and sharing of information on and assessments of emerging scientific and technological innovations, beginning at the earliest possible stage of their development;

[38] On the benefits of the framework convention-protocol approach, see id.; Abbott, Kenneth W., and Duncan Snidal. 2004. Pathways to international cooperation. In *The impact of international law on international cooperation,* eds. E. Benvenisti and M. Hirsch; Abbott, Kenneth W., Gary Marchant, and Douglas Sylvester. 2006. A framework convention for nanotechnology? *Environmental Law Reporter News and Analysis* 36: 10931.

ii. cooperate in good faith to coordinate national regulatory actions (broadly defined), while recognizing that such actions may vary due to differing regulatory cultures and legal systems;
iii. involve technical experts, researchers, workers, consumers and other stakeholders from the private sector and civil society, as appropriate, in assessing emerging innovations and framing and evaluating regulatory responses;
iv. adopt, and cooperate to promote efficient, effective and appropriate forms of regulation that do not unduly hamper innovation or impose excessive costs, while protecting the public;
v. minimize the impact of national regulatory actions on international trade and other forms of international economic activity[39];
vi. take into account the special needs and circumstances of developing countries and economies in transition, in accordance with the principle of common but differentiated responsibilities; and
vii. (perhaps most controversially) apply some form of the precautionary principle in dealing with potentially serious risks of early-stage innovations.[40]

In addition, the agreement might include certain *general commitments,* especially regarding research and information sharing, collaboration and the general character of national regulatory procedures. Again drawing on Vienna and the FCCC, the parties to FASTIR should commit themselves to:

i. promote and cooperate in scientific, technological, socio-economic, ethical and other forms of research on and assessment of emerging innovations, beginning at the earliest possible stage in their development, with the aim of reducing uncertainty as to their potential benefits, risks and social, economic and ethical implications[41];
ii. cooperate to fully, openly and promptly exchange information on and assessments of emerging innovations and their potential benefits, risks and other impacts (subject to protections for trade secrets and other sensitive private information), through mechanisms established under the convention and directly among themselves[42];

[39]The WTO Agreements on Technical Barriers to Trade (TBT) and on the Application of Sanitary and Phytosanitary Measures (SPS) include useful formulations of such principles. For example, the TBT Agreement provides that (a) technical regulations should not "create unnecessary obstacles to international trade;" and that (b) while states should be free to adopt necessary regulations, measures should not be applied in ways that constitute "arbitrary or unjustifiable discrimination" between states or "a disguised restriction on international trade." TBT Agreement, Preamble, para. 6.

[40]See, e.g., FCCC Art. 3:3: "The Parties should take precautionary measures to anticipate, prevent or minimize the causes of climate change and mitigate its adverse effects. Where there are threats of serious or irreversible damage, lack of full scientific certainty should not be used as a reason for postponing such measures, taking into account that policies and measures . . . should be cost-effective so as to ensure global benefits at the lowest possible cost."

[41]Cf. FCCC Art. 4:g; Vienna Art. 2:2(a).

[42]Cf. FCCC Art. 4:h.

iii. provide technical and financial assistance, individually and collectively, to developing country parties to assist them in gathering information on and assessing the benefits, risks and other impacts of innovations, sharing such information and assessments, and formulating legal and policy responses;
iv. evaluate the environmental, health and safety risks and the social and ethical implications of scientific and technological innovations along with their benefits;
v. shape national regulatory processes to facilitate the coordination of legal and policy responses to emerging innovations, subject to variations due to differing regulatory cultures and legal systems, and to share information on the consequences of various response strategies[43];
vi. give adequate advance notice to other parties of legal and policy actions that might have adverse external consequences, e.g., on international trade, through mechanisms established under the convention, and consult in good faith regarding any such consequences; and
vii. participate in good faith in the institutions and procedures the convention creates.

9.4.4 Host Institution

The heart of FASTIR would be the creation of institutions and procedures for the coordination, steering and facilitation of national regulatory actions. As treaty bodies, the core institutions would be inter-state in nature.

The initial structural decision would be the selection of an organization to host FASTIR and provide the administrative, financial and intellectual services needed to build and operate an effective operational regime. The convention could create a new, freestanding host institution, but for reasons of cost and efficiency it would be preferable to identify an existing organization that provides a good "fit" with the objectives and mechanisms of the agreement.

"Fit" turns on three major factors. First, membership in the host organization should be roughly congruent with the states that are expected or desired to participate in the convention; at the least, all the essential convention parties should be members of the organization. This is often a difficult problem for treaty negotiators, as the international system includes a limited supply of effective organizations from which to choose. If perfect congruence cannot be achieved, as is often the case, the convention can provide special arrangements for collaboration by non-members. Second, the host organization should have the authority, expertise and experience to engage in the legal and political working methods envisioned by the convention. For example, if the convention contemplates the adoption of legally binding protocols, the host organization should be authorized to develop binding instruments and have

[43]Cf. FCCC Art. 4:g-h.

experience in doing so; it should not be a purely soft law institution.[44] Third, the fundamental principles of the host organization should be consistent with those of the convention.

The UN specialized agency with the most specific science policy mandate is UNESCO – the United Nations Educational, Scientific and Cultural Organization.[45] UNESCO has certain advantages as a host institution. It is a true multilateral organization, with over 190 members, so all FASTIR parties are almost certain to belong. It has adopted both treaties, such as the Convention Concerning the Protection of the World Cultural and Natural Heritage (1972), and soft law instruments, such as the Universal Declaration on Bioethics and Human Rights (2005) and the International Declaration on Human Genetic Data (2003). Its principles emphasize the exchange of information, capacity building in developing countries, and respect for individual cultures, all likely to be consistent with FASTIR.

Yet UNESCO also has significant weaknesses. First, it focuses primarily on the cultural, human rights and related aspects of scientific issues within its mandate, as reflected in its Declaration on Bioethics and Human Rights and its Universal Declaration on the Human Genome and Human Rights (1997). As a result, while UNESCO could add substantial value in assessing the social, cultural and ethical impacts of innovations, it may be too narrow to serve as a host organization. Second, UNESCO's current priorities are to build scientific and science policy capacities in developing countries, while promoting poverty alleviation and sustainable development.[46] It might be an uncomfortable fit for the industrial countries that dominate scientific innovation. Third, some UNESCO policies have been highly controversial: the US, UK and Singapore withdrew from the organization in the 1980s, and the US remained outside for nearly 20 years.[47] Recent reforms have moderated these controversies, but UNESCO may still retain unnecessary political "baggage."

The OECD may be a more suitable host, although it too has weaknesses. The members of the organization are the 30 leading industrialized states, so most essential FASTIR parties – the states that are most active in advanced scientific and technological research and development – are OECD members. To be sure, some states important for the convention – notably the BRIC countries[48] and other developing and transitional economies where research is developing rapidly – are not

[44]The converse is also true, although that situation is less common: if the convention contemplates operating through soft law, it should not affiliate with an organization (such as the WTO) that focuses solely on hard law. See Abbott and Snidal, Pathways, supra note 38.

[45]See http://portal.unesco.org/en/ev.php-URL_ID=3328&URL_DO=DO_TOPIC&URL_SECTION=201.html; http://portal.unesco.org/science/en/ev.php-URL_ID=5802&URL_DO=DO_TOPIC&URL_SECTION=201.html, visited 10 Sept 2009.

[46] http://portal.unesco.org/science/en/ev.php-URL_ID=5805&URL_DO=DO_TOPIC&URL_SECTION=201.html, visited 10 Sept 2009. A leading priority is gender equality. http://portal.unesco.org/science/en/ev.php-URL_ID=5157&URL_DO=DO_TOPIC&URL_SECTION=201.html, visited 10 Sept 2009.

[47]See http://portal.unesco.org/en/ev.php-URL_ID=14606&URL_DO=DO_TOPIC&URL_SECTION=201.html, visited 10 Sept 2009.

[48]Brazil, Russia, India and China.

OECD members. In 2007, however, the OECD began a significant enlargement, inviting Russia as well as Chile, Estonia, Israel and Slovenia to begin discussions on membership and adopting a "road map" for negotiating their accession. It also offered "enhanced engagement, with a view to possible membership," to Brazil, China, India, Indonesia and South Africa.[49] Some of these countries are also observers in OECD working parties that are considering specific emerging technologies. In addition, the OECD operates extensive outreach programs through its Centre for Co-operation with Non-Members. It sponsors ten "global forums" including non-members that address transnational problems; two new global forums, on biotechnology and the knowledge economy, are currently being created. Nonetheless, as critics such as the ETC Group argue, the structure of the OECD does not fully incorporate the views of developing countries that would feel the effects of any environmental, economic or other adverse consequences created by risky innovations.[50] To address this problem, FASTIR should provide adequate mechanisms for participation and input from concerned states not members of the OECD.

OECD working methods are also suitable. Most generally, the organization's work is driven by research and analysis[51]; it has a large expert staff, and is among the world's leading sources of economic and social data and forecasts as well as analyses of economic and social policy. National government experts provide, review and disseminate most of this information. The OECD is also familiar with transgovernmental and transnational policy-making. Most of its work is done in some 200 specialized committees and expert groups, in which representatives of national government agencies, staff and external experts share policy experiences, lessons learned and best practices, aiming for policy coordination.[52] It also receives regular input from the private sector and workers through the Business and Industry Advisory Committee (BIAC) and the Trade Union Advisory Committee (TUAC), as well as other civil society links.[53] Finally, while the OECD relies primarily on soft law, including model treaties, best practice recommendations and guidelines, it also adopts and implements binding legal instruments, including conventions[54] and mandatory decisions.[55]

---

[49] "OECD invites five countries to membership talks, offers enhanced engagement to other big players," 16 May 2007, available at http://www.oecd.org/document/33/0,3343,en_2649_34487_38603809_1_1_1_1,00.html, visited 10 Sept 2009.

[50] ETC Group, Nanogeopolitics 2009: The Second Survey (draft July 2009), at 1, 10, http://www.lawbc.com/other_pdfs/00048599.PDF, visited 7 Sept 2009.

[51] See http://www.oecd.org/pages/0,3417,en_36734052_36761681_1_1_1_1_1,00.html, visited 7 Sept 2009.

[52] See http://www.oecd.org/pages/0,3417,en_36734052_36761791_1_1_1_1_1,00.html, visited 7 Sept 2009.

[53] See http://www.oecd.org/pages/0,3417,en_36734052_36761800_1_1_1_1_1,00.html, visited 7 Sept 2009.

[54] For example, the Convention on Combating Bribery of Foreign Public Officials in International Business Transactions.

[55] For example, the Code of Liberalisation of Capital Movements, adopted as a binding decision of the OECD Council.

Finally, the OECD already addresses significant aspects of science and technology policy,[56] including innovation,[57] international research collaboration,[58] nanotechnology[59] and biotechnology[60]; the fundamental principles it follows in these fields are largely consistent with those animating FASTIR. To be sure, the economic and market orientation of the organization raises a note of caution. As noted above, OECD work on scientific and technological innovation has focused on developing incentives and appropriate market structures for innovation, access to global markets for products of innovation, and similar issues, as well as enhancing public support for R&D; in its work on innovation, the OECD has devoted little attention to the regulation of risks and other adverse impacts.[61] In terms of regulatory policy, moreover, the OECD has been at the forefront of efforts to streamline regulation,[62] promote regulatory impact assessment procedures,[63] and reduce administrative burdens.[64] However, the same OECD department that addresses regulatory policy has also considered risk regulation, and its work in that area has strived for balance.[65] Given its other advantages, the economic orientation of the OECD should not disqualify it as host for FASTIR.

9.4.5 Treaty Institutions

FASTIR would establish treaty bodies through which the state parties could take action under the convention. I merely sketch these institutions here, as their specific structures and functions would derive from the activities assigned to them.

The core institution of a framework convention is a *Conference of the Parties (COP)*. The COP is not simply the state parties meeting as a group, but an institution with its own authorities and decision-making procedures. The COP is typically authorized to promote agreed actions, such as information exchange and regulatory coordination. In addition, as the ultimate authority under the convention, the

56 See http://www.oecd.org/department/0,3355,en_2649_34269_1_1_1_1_1,00.html, visited 7 Sept 2009.

57 See note [33] supra.

58 http://www.oecd.org/department/0,3355,en_2649_34319_1_1_1_1_1,00.html, visited 7 Sept 2009.

59 http://www.oecd.org/site/0,3407,en_21571361_41212117_1_1_1_1_1,00.html, visited 7 Sept 2009.

60 http://www.oecd.org/department/0,3355,en_2649_34537_1_1_1_1_1,00.html, visited 7 Sept 2009.

61 http://www.oecd.org/document/50/0,3343,en_41462537_41454856_41488882_1_1_1_1,00.html, visited 10 Sept 2009.

62 See note 24 supra and accompanying text.

63 Ibid.

64 http://www.oecd.org/document/43/0,3343,en_2649_34141_38227179_1_1_1_1,00.html, visited 7 Sept 2009.

65 http://www.oecd.org/document/23/0,3343,en_2649_34141_37551127_1_1_1_1,00.html, visited 7 Sept 2009.

COP adopts budgets, procedural rules and reports, establishes subsidiary bodies and makes similar administrative decisions. Most importantly, the COP is charged with continuously reviewing the adequacy of the convention's rules and procedures in light of advancing scientific knowledge, information on party compliance, and evidence of the convention's effectiveness. If the convention is judged to be insufficient in any of these areas, the COP may recommend – or even adopt – modifications.

The COP is supported by specialized committees and other *subsidiary bodies*. These may include traditional standing committees such as budget and administration; committees for particular fields of innovation, such as nanotechnology or synthetic biology; and a committee to supervise implementation and compliance.[66] Perhaps most significantly, modern conventions typically establish a scientific and technical committee to advise the COP, coordinate research collaboration, promote technical activities by participating states, assess the impact of the convention, and undertake similar important tasks.[67] If states adopt specific protocols under the framework convention, each would create its own Meeting of the Parties (MOP), analogous to COP for the parties to that protocol; a protocol can create its own subsidiary bodies or share those of the convention. Finally, the convention normally establishes a *secretariat,* which may be the host institution or may be housed there. The secretariat is responsible for day-to-day administration, but may also have considerable substantive influence, e.g., by analyzing information, consulting with experts and civil society, proposing modifications, promoting compliance, and controlling technical and financial assistance.

Finally, FASTIR should provide for *relations with other institutions*. Institutional relationships are particularly important in this case, for two reasons. First, FASTIR would rely on government agencies and officials and on a range of non-governmental actors (e.g., researchers, business firms, civil society organizations) to carry out much of its work. The convention should authorize relationships with such actors and organizations, and direct the COP and secretariat to establish and manage those relationships. Second, FASTIR – and hence the range of potential protocols – would have a very broad scope, extending over many aspects of scientific and technological innovation and regulation. It therefore faces a substantial risk of overlap or even conflict with other international agreements: many environmental treaties, for example, provide for regulation of technologies, scientific and technological research, technology transfer, information sharing and related matters.

One approach to institutional relationships – widely followed, but formal and narrow – is to authorize relevant international organizations and treaty secretariats, as well as qualified non-governmental organizations, to participate as "observers"

[66]For example, FCCC Art. 10, "Subsidiary Body for Implementation."

[67]For example, FCCC Art. 9, "Subsidiary Body for Scientific and Technological Advice." Some organizations with functions similar to those of FASTIR have elaborate structures of advisory and operational committees and other bodies. An interesting example is ICANN, the Internet Corporation for Assigned Names and Numbers. See http://www.icann.org/en/structure/, visited 9 Sept 2009.

in meetings under the convention.[68] Because FASTIR would initiate a high level of trans-governmental and transnational engagement, however, an approach that enables more extensive, informal cooperation would be more valuable. For example, the Convention on Biological Diversity (a similarly broad agreement which operates to some extent as a framework convention) directs its COP to contact the institutions of relevant treaties "with a view to establishing appropriate forms of cooperation with them."[69] Even more broadly, the FCCC directs its COP to "seek and utilize, where appropriate, the services and cooperation of, and information provided by, competent international organizations and intergovernmental and non-governmental bodies."[70]

9.4.6 Expedited Procedures

Finally, as a crucial part of the framework convention-protocol approach, FASTIR would establish procedures for collective action when the convention parties determine that action is required in a particular area. As noted earlier, this might be a specific field of scientific or technological research; a group of such fields; or one or more cross-cutting issues relevant to many areas of scientific and technological innovation.

Initially, the convention would have to establish procedures by which the state parties could identify and define specific areas or issues on which action should be taken. A wide range of approaches is possible. At one extreme, the FCCC prescribes only minimal procedures for initiating rule-making: any party may propose an amendment for action by the COP; the COP may adopt any protocol it wishes.[71] In practice, the COP has used its broad authority to create a rule-making process, still highly flexible, managed by the COP Bureau (akin to an executive committee) with support from the secretariat and advice from specialized committees.[72] At the other extreme, the International Labor Organization (ILO) administers a complex, two-year ("double discussion") legislative process that involves several ILO bodies and requires consultations with governments, workers and employers.[73] For FASTIR, the general approach of the FCCC, including authority for the COP to develop detailed procedures, might be best.

---

[68] For example, UNFCCC, Art. 7:6.

[69] Convention on Biological Diversity, Art. 23:h.

[70] FCCC, Art. 7:1.

[71] UNFCCC, Art. 15–17.

[72] http://unfccc.int/essential_background/convention/convention_bodies/bureau/items/3431.php, visited 8 Sept 2009.

[73] http://www.ilo.org/global/What_we_do/InternationalLabourStandards/Introduction/creation/lang–en/index.htm; International Labor Standards Handbook of Procedures, section I, http://www.ilo.org/ilolex/cgi-lex/pdconv.pl?host?status01&textbase=iloeng&document=18&chapter=29&query=(%23docno%3D25200602A)+%40ref&hightlight=&querytype=bool&context=0, visited 8 Sept 2009.

The convention would then have to prescribe procedures for final action. Ideally these would be faster and less costly than the interstate negotiations needed to create a new treaty; in particular, Vienna and the FCCC each authorize the COP to take action by voting rather than by more time-consuming negotiation and agreement. The COP might act by adopting "protocols" – such as the well-known Montreal and Kyoto protocols – which set their own requirements for entry into force (e.g., minimum number of parties, national formalities). Vienna and the FCCC also authorize the COP to adopt amendments to the convention, "annexes" containing scientific, technical and administrative information, and amendments to protocols and annexes. The ozone regime in particular has made ample use of these procedures, repeatedly expanding its coverage of ozone-depleting substances and tightening its production and consumption limits and phase-out schedules as new information has become available.

In the remainder of this chapter, I sketch less formal modes of action the convention might also establish.

9.5 Information

For states to deal effectively with rapidly emerging innovations, they must establish processes that guarantee the production and sharing of early stage information on innovations and their potential benefits, risks and other impacts. In addition, to deal with the three types of uncertainty described above, states must have mechanisms to assess information as it emerges – in a technical sense, an economic, social and political sense, and often a normative sense. Even if we focus only on international efforts to coordinate, steer and facilitate national processes and to share the results across borders, these are challenging problems requiring institutional creativity. The ultimate solution – both domestically and internationally – will almost certainly require the engagement of all three levels of governance.

At the *transnational* level, FASTIR might first encourage states to establish domestic procedures that engage their scientific and technical communities, private sector and civil society in producing information, assessing it from multiple perspectives, and sharing the results.[74] It should also encourage them to adopt at least roughly similar procedures, subject to the many social, political and other differences among them, and to develop and use harmonized nomenclatures, indicators and other analytical tools. This form of steering could be accomplished through the adoption of principles or commitments, discussions in the COP or relevant committees, or efforts by the secretariat. The FASTIR parties could also fund technical assistance to help states with limited capacity to design and implement such procedures.

[74] The 1998 Aarhus Convention on Access to Information, Public Participation in Decision-Making and Access to Justice in Environmental Matters provides for rights of public participation in certain types of governmental decisions relating to the environment, and might be relevant to the procedures discussed here in states that have ratified that convention. See http://www.unece.org/env/pp/welcome.html, visited 10 Sept 2009.

FASTIR should also establish its own transnational information procedures. The core function of these mechanisms would be to encourage and promote the sharing of information across borders. Information sharing alone, however, would leave the assessment of innovations to individual states.[75] On the theory that multiple perspectives lead to better informed assessments,[76] FASTIR should go further, aiming to produce truly transnational assessments. For example, its scientific and technical advisory body could be tasked not only with traditional "vertical" functions such as reviewing the effectiveness of the convention and advising the COP, but also with "horizontal" functions such as sharing information and formulating common technical assessments, drawing on participating experts and the scientific communities they represent. FASTIR might establish similar transnational advisory bodies that represent technologically sophisticated business firms and civil society stakeholders, including workers, consumers and other concerned groups. Each body could share information and formulate collective assessments as well as advising the COP. Assessing social, cultural and ethical implications would be particularly challenging, as it requires normative deliberation; a multi-stakeholder body, broadly representative of societies and cultures, that brings together scientific and technological innovators and the stakeholders likely to be affected by their innovations would be best positioned to carry out this responsibility.

This level of private engagement would be unusual in an international institution, but it is essential in a regime devoted to ameliorating diverse forms of uncertainty. And it is not unprecedented. A similar approach has characterized the "transatlantic regulatory cooperation" (TRC) arrangements established by the US and the EU since the mid-1990s.[77] In addition to promoting cooperation among government agencies, TRC established people-to-people "dialogues," spanning the two polities, to advise government negotiators on regulatory issues. The most influential has been the Transatlantic Business Dialogue[78]; its views have largely set the TRC agenda. Others include the Transatlantic Consumer Dialogue[79] and a now nearly moribund Transatlantic Labor Dialogue. These Dialogues have somewhat different purposes than the bodies suggested here, but they indicate that it is feasible to incorporate non-state actors into international regimes.

Trans-governmental networks would be effective ways to share information, not only on innovations, but also on government interventions attempted or contemplated; consultations among officials could produce better informed regulation and avoid or address many international regulatory externalities. Trans-governmental

[75]A commitment to share information and assessments was suggested above as an element of the convention.

[76]See p. 130, supra.

[77]For a history of these efforts, see Ahearn, Raymond J. Congressional research service, CRS report for congress – Transatlantic regulatory cooperation: Background and analysis, Oct. 22, 2008, available at http://fas.org/sgp/crs/misc/RL34717.pdf, visited 10 Sept 2009.

[78]http://www.tabd.com/, visited 10 Sept 2009.

[79]http://tacd.org/index.php?option=com_frontpage&Itemid=1, visited 10 Sept 2009.

networks would be equally effective at harmonizing nomenclatures and methodologies. They would allow regulators to share lessons learned on issues such as procedures for gathering early stage information on innovations, the conduct of technology assessments, the management of stakeholder and public input, and other governance issues. They could even encourage regulators to share and compare their assessments of risks and impacts, with the possibility of reformulating them based on peer input.

US-EU TRC programs have relied predominantly on trans-governmental approaches. For example, the 2005 Roadmap for US-EU Regulatory Cooperation[80] provided explicitly for collaboration on sectoral issues among agencies such as the US FDA, the EU DG Enterprise and Industry Pharmaceuticals Unit, and the European Medicines Agency; the US National Highway Traffic Safety Administration and the DG Enterprise and Industry Automobile Unit; and the US Consumer Product Safety Commission and DG Health and Consumer Affairs.[81] FASTIR could either establish permanent networks of regulators from the most relevant agencies or could establish a mechanism for creating ad hoc networks as issues arose.

Finally, at the *inter-state* level, international agreements contain many procedures for information-sharing and collaboration that can serve as models. The FCCC provides a good illustration.[82] This framework convention commits all state parties to develop, publish and share information on their greenhouse gas sources and sinks (using comparable methodologies developed by the COP) and on national mitigation measures; it further commits them to cooperate in scientific, technological, socio-economic and other research and to fully, openly and promptly exchange information.[83] Parties agree to support international research programs and networks, and to strengthen national research capabilities.[84] And the FCCC directs the COP to facilitate information exchange, coordinate national measures and develop comparable methodologies.[85]

---

[80]http://ec.europa.eu/enterprise/regulation/better_regulation/docs/docs_conference_EU_US_260106/2005_roadmap.pdf, visited 10 Sept 2009.

[81]For a recent summary of progress in these and similar initiatives, see Transatlantic Economic Council Report to the EU-US Summit 2008, available at http://www.eurunion.org/partner/summit/Summit2008-06-10/2008EU-USSummitDeclar-6-10-08.pdf#page=16, visited 10 Sept 2009.

[82]In addition to the provisions discussed here, the FCCC directs its secretariat to cooperate with the Intergovernmental Panel on Climate Change (IPCC), a body established by the World Meteorological Organization and UN Environment Programme in 1989. The IPCC provides comprehensive periodic reviews of the state of scientific knowledge on climate change, its impacts and possible responses; it functions as an independent international source of information and assessment. This chapter, however, considers only mechanisms to encourage national governments and agencies to produce and share information.

[83]FCCC Art. 4:1. See also Art. 12.

[84]Ibid., Art. 5.

[85]Ibid., Art 7.

9.6 Steering National Regulatory Systems

As a framework convention, FASTIR would not set the terms for the regulation of specific technologies or risks; that would be the function of subsequent protocols. Nonetheless, FASTIR could include an important, though more general, substantive component: principles and mechanisms designed to steer national institutions and procedures toward desirable *approaches* to technology regulation, and to facilitate implementation of those approaches in states with limited regulatory capacity. Agreed principles and procedures would also make it easier to coordinate national regulatory actions.

9.6.1 Science, Technology and Policy-Making

A major focus of these efforts should be the production and handling of scientific and technological information relevant to regulation. Especially when innovation is dynamic, effective and legitimate regulation requires institutions and processes that provide timely and appropriate scientific input into policy-making and that guarantee the integrity of that input. Currently, even some advanced states lack well-developed institutions and processes for scientific policy-making, while those that exist are far from comparable. Here again, modest international efforts could produce significant benefits.

In terms of institutions, for example, the United Kingdom Parliament is advised by the Parliamentary Office of Science and Technology (POST), which aims to provide "independent, balanced and accessible analysis of public policy issues related to science and technology."[86] POST also engages in "horizon-scanning to anticipate issues of science and technology that are likely to impact on policy."[87] POST and similar organizations advising other European legislatures, including the European Parliament, have formed the European Parliamentary Technology Assessment (EPTA) network "as an aid to the democratic control of scientific and technological innovations." In addition to enhancing each member's own work, EPTA can undertake joint technology assessments.[88] In the US, in contrast, although Congress created the Office of Technology Assessment (OTA) in 1972 to provide "objective and authoritative analysis of complex scientific and technical issues to aid in policymaking,"[89] it terminated funding for OTA in 1995.[90]

---

[86] http://www.parliament.uk/parliamentary_offices/post.cfm, visited 10 Sept 2009.

[87] Ibid.

[88] http://www.eptanetwork.org/EPTA/about.php, visited 10 Sept 2009.

[89] Knezo, Genevieve. Congressional research service, "Technology assessment in congress: History and legislative options," CRS report for congress RS21586, May 20, 2005, at 1. OTA is explicitly credited as the inspiration for EPTA. http://www.eptanetwork.org/EPTA/about.php, visited 10 Sept 2009.

[90] Knezo, supra note 89, at 1. The Government Accountability Office (GAO) has been granted temporary authority to conduct technology assessments on a pilot basis, and proposals to make

In the executive branch, both the UK and the US have government science advisers: in the UK, the Government Chief Scientific Adviser (GCSA) in the Government Office for Science[91]; in the US, the Director of the Office of Science and Technology Policy (OSTP) in the Executive Office of the President.[92] Here too, however, the UK has introduced potentially worthwhile innovations. For example, since 2002 it has created a network of Chief Scientific Advisers based in a wide range of executive departments; each Adviser is charged with supporting the use of scientific evidence in policy-making, ensuring the quality of scientific inputs, and making its department a better consumer of science.[93] In addition, the government has adopted a common code of practice for scientific advisory committees[94] as well as common guidelines for executive departments and agencies on the use of science in policy-making.[95]

In terms of process, perhaps the most significant issue is guaranteeing integrity in the supply and use of scientific and technological information. In the US during the administration of President George W. Bush, many scientists and other critics challenged the integrity of the policy process, charging that scientific input was manipulated for political ends.[96] Such interference must clearly be avoided if emerging innovations are to be properly understood and regulated. President Barack Obama has responded strongly to these critiques, charging the Director of OSTP with "the responsibility for ensuring the highest level of integrity in all aspects of the executive branch's involvement with scientific and technological processes."[97] More concretely, the President has instructed the Director to "develop recommendations for Presidential action designed to guarantee scientific integrity throughout the executive branch;" those recommendations are to reflect stated principles, including

---

that arrangement permanent or to create an agency similar to OTA have been regularly introduced. Ibid., at 3–6.

[91] http://www.chiefscientificadviser.ie/, visited 10 Sept 2009

[92] http://www.ostp.gov/, visited 10 Sept 2009.

[93] House of Commons, Science and Technology Committee. "Scientific advice, risk and evidence based policy making: Government response to the committee's seventh report of session 2005–06," First special report of session 2006–07, HC 307, Feb. 27, 2007, at 2.

[94] Government Office for Science. Code of practice for Scientific Advisory Committees, December 2007, available at http://www.dius.gov.uk/partner_organisations/office_for_science/science_in_government/strategy_and_guidance/~/media/publications/F/file42780, visited 10 Sept 2009

[95] HM Government. Guidelines on scientific analysis in policy making, October 2005, available at http://www.berr.gov.uk/files/file9767.pdf, visited 10 Sept 2009

[96] See, e.g., 2004 Union of concerned scientists statement on restoring scientific integrity to federal policy making, available at http://www.ucsusa.org/scientific_integrity/abuses_of_science/scientists-sign-on-statement.html, visited 10 Sept 2009; 2008 Union of concerned scientists statement on scientific freedom and the public good, available at http://www.ucsusa.org/scientific_integrity/abuses_of_science/scientific-freedom-and-the.html, visited 10 Sept 2009.

[97] Memorandum for the Heads of Executive Departments and Agencies, "Scientific integrity," March 9, 2009, available at http://www.whitehouse.gov/the_press_office/Memorandum-for-the-Heads-of-Executive-Departments-and-Agencies-3-9-09/, visited 10 Sept 2009.

the appointment of science and technology officials based on expertise and integrity rather than political considerations, and reliance on scientific processes such as peer review.[98]

International cooperation on these issues has been very limited. The OECD regularly considers a variety of science and technology policy issues, including innovation as a spur to economic growth,[99] public sector research,[100] and international cooperation in basic research[101]; but it has not attempted to steer or coordinate national policy in the areas considered here. FASTIR could make an important contribution by enunciating principles and developing recommendations, best practices, guidelines, or even binding rules on the sources of scientific and technological information and analysis, the integrity of that information, and its appropriate use by legislators, regulatory agencies and executive policy-makers.

9.6.2 Regulatory Practice

A second focus area should be the promotion of good regulatory practices. Steering and facilitation in this area would encourage more efficient, effective and legitimate responses to innovations, and would increase the harmonization of national regulatory procedures and techniques, easing coordination of specific regulatory actions. By improving poor regulatory practices in particular states, these efforts could combat the emergence (or persistence) of "risk havens" – states with sufficient scientific and technological capacity to produce or consume risky innovations, but without the governance capacity to regulate appropriately – as well as the broader competitive incentives to under-regulate.

FASTIR could draw from existing international programs on regulatory practice. One useful building block might be the "Better Regulation" strategy implemented by the European Commission since 2002, expanded in 2005 as part of the revised "Lisbon Strategy" to stimulate economic growth and employment.[102] Better Regulation is said to be motivated in part by the need to respond to rapid technological change[103]; however, its principal goal appears to be reducing the economic burdens of regulation.

---

[98]Ibid.

[99]http://www.oecd.org/document/7/0,3343,en_2649_34273_1911303_1_1_1_1,00.html, visited 10 Sept 2009.

[100]http://www.oecd.org/about/0,3347,en_2649_34293_1_1_1_1_1,00.html, visited 10 Sept 2009.

[101]http://www.oecd.org/department/0,3355,en_2649_34319_1_1_1_1_1,00.html, visited 10 Sept 2009.

[102]Communication from the Commission to the Council and the European Parliament – Better Regulation for Growth and Jobs in the European Union, COM/2005/0097 final, 16 March 2005.

[103]European Commission, Better Regulation – Simply Explained (2006), at 4, available at http://ec.europa.eu/governance/better_regulation/documents/brochure/br_brochure_en.pdf, visited 10 Sept 2009.

Better Regulation has five major components: (1) assessing the economic, social and environmental impacts of proposed regulations (regulatory impact assessment)[104]; (2) assessing the "red tape" and other administrative burdens of proposed regulations; (3) consulting with stakeholders; (4) considering alternatives to mandatory regulation, including non-binding recommendations, directives that allow national flexibility rather than uniform regulations, co-regulation (entrusting the achievement of regulatory goals to private parties), and self-regulation; and (5) simplifying existing regulations.[105] The Commission applies these measures at EU level and encourages member states to do so domestically.

Better Regulation clearly resembles the US regulatory review procedure conducted since 1981 by the Office of Management and Budget (OMB), first under President Reagan's Executive Orders 12291 and 12498, and since 1993 under President Clinton's Executive Order 12866,[106] as well as various OMB guidance instruments.[107] However, the European Commission notes two significant differences: OMB has focused heavily on cost-benefit analysis of proposed regulations, whereas Better Regulation calls for a broader analysis of alternatives; and OMB considers only administrative regulations, whereas Better Regulation also covers European measures equivalent to legislation.[108] On the first point, the Obama administration may again change course: it has directed OMB to make recommendations for a new executive order on regulatory review; those recommendations are to address elements such as "the role of cost-benefit analysis; . . . distributional considerations, fairness, and concern for the interests of future generations; . . . methods of ensuring that regulatory review does not produce undue delay; . . . [and] the role of the behavioral sciences . . . ."[109]

Two other existing programs are also relevant. First, since 1995 the OECD has developed principles for good regulation and regulatory reform; it applies those principles to its member states through peer review and disseminates them to non-members.[110] The current Guiding Principles for Regulatory Quality and Performance were adopted in 2005.[111] They emphasize competition, efficiency, deregulation and open trade, with the goal of enhancing economic growth and productivity, and call for regulatory impact assessment. As noted above, the

---

[104]The economic aspect was strengthened in 2005. See Communication, supra note 102, at section 2:A.

[105]See Communication, supra note 102.

[106]E.O. 12866, "Regulatory Planning and Review," 30 Sept 1993, 58 Fed. Reg. 51735, 4 Oct 1993.

[107]For example, Office of management and budget, circular A-4, "Regulatory analysis," 17 Sept 2003, available at http://www.whitehouse.gov/omb/assets/regulatory_matters_pdf/a-4.pdf, visited 10 Sept 2009.

[108]Communication, supra note 102, at 15. On cost-benefit analysis, see Circular A-4, supra note 107, at 2–3.

[109]Memorandum for the Heads of Executive Departments and Agencies, "Regulatory Review," Jan. 30, 2009, 74 Fed. Reg. 5977, 3 Feb 2009.

[110]http://www.oecd.org/about/0,3347,en_2649_34141_1_1_1_1_1,00.html, visited 10 Sept 2009.

[111]http://www.oecd.org/dataoecd/24/6/34976533.pdf, visited 10 Sept 2009.

OECD has also considered issues of risk assessment and management.[112] Second, US-EU TRC programs have focused on avoiding barriers to trade created by disparate regulatory processes, especially product standards, testing and certification.[113] The 2002 Guidelines on Regulatory Cooperation and Transparency[114] encourage regulators to consult before such regulations are adopted.[115] The 2005 Roadmap[116] establishes dialogues among regulators on specific products, as well as cross-cutting dialogues. An OMB-European Commission dialogue has helped the agencies understand their respective approaches to regulatory review, but has produced little harmonization.[117]

Unfortunately, while all of these programs help to constrain excessive regulation and adverse impacts on trade, and some, such as Better Regulation, encourage flexible alternatives to command-and-control regulation, none of them seems well designed for the challenges posed by dynamic scientific innovation. National programs like Better Regulation and OMB review, moreover, differ in significant ways. Thus, while FASTIR could build on these programs, it would be necessary to modify them to address the unique problems of dynamic innovation: e.g., by encouraging the appropriate use of scientific information and assessments; ensuring balanced consideration of potential benefits, risks and other impacts along with regulatory burdens; and providing tools for rapid action in case of emerging threats.

9.6.3 Trans-governmental Dialogue

Both science and technology policy and good regulation are appropriate areas for a trans-governmental approach. Beyond traditional regulatory bodies such as environment or food and drug agencies, trans-governmental arrangements could be extended to legislators and to specialized actors such as science advisors and members of scientific advisory committees.

Many of the international norms and programs on science and technology policy and regulation just discussed have been created through trans-governmental

[112]http://www.oecd.org/document/23/0,3343,en_2649_34141_37551127_1_1_1_1,00.html, visited 10 Sept 2009.

[113]A major goal of TRC was to negotiate mutual recognition agreements that would obviate disparate product regulation. However, only three sectoral agreements were adopted. See Ahearn, CRS Report, supra note 77.

[114]http://ec.europa.eu/enterprise/regulation/better_regulation/docs/docs_conference_EU_US_260106/GUIDELINES_EU-US_FINAL.pdf, visited 10 Sept 2009.

[115]The 2006 Best Cooperative Practices distill lessons for such consultations. http://trade.ec.europa.eu/doclib/docs/2006/july/tradoc_129223.pdf, visited 10 Sept 2009. In spite of TRC, some major regulations, such as the EU REACH program for chemicals, were adopted without the contemplated notice and consultation. Ahearn, supra note [77].

[116]Note 80 supra.

[117]See http://www.whitehouse.gov/omb/assets/regulatory_matters_pdf/draft_sg-omb.pdf, visited 10 Sept 2009.

deliberations – e.g., in the OECD Working Party on Regulatory Management and Reform and Working Group on Technology and Innovation Policy. Others rely on trans-governmental deliberations to achieve their ends – e.g., through the sectoral US-EU regulatory dialogues and the cross-cutting OMB-Commission dialogue. However, virtually all of these bodies are composed of officials with a single mandate and viewpoint. Thus, the OECD Working Party is made up of "officials responsible for cross-cutting and horizontal regulatory reform policies,"[118] while its Working Group is composed of "officials responsible for science, technology and innovation."[119] Such institutions are likely to produce understandings skewed to their particular interests and modes of thinking, without adequate consideration of competing concerns and approaches. FASTIR could make an important contribution by establishing balanced trans-governmental bodies that could consider potential tradeoffs among regulatory approaches; identify best practices in the use of scientific information for policy-making and in appropriate, timely and effective regulation; promote those practices through broad consultations, technical assistance and peer review; and facilitate their widespread adoption by governments.

9.7 Regulatory Action

When it becomes necessary to coordinate concrete regulatory measures (broadly defined) aimed at particular innovations, risks or other impacts, all three levels of governance should once again be engaged.

At the *transnational* level, FASTIR could follow an approach a co-author and I call "transnational new governance:" promoting and orchestrating the development and implementation of self-regulation and voluntary codes of conduct among the private actors responsible for developing and commercializing scientific and technological innovations, alone and in cooperation with concerned stakeholders. This approach would draw on the knowledge and capacities of those actors, making them part of the overall regulatory system, not mere targets of regulation.[120]

As Brian Rappert's chapter makes clear, professional codes of conduct have for the most part failed to address regulatory issues such as those considered here, and have been weakly implemented and enforced.[121] Outside of the professions, however, considerable progress toward self-regulation and voluntary multi-stakeholder codes has already been made, especially in industries and areas of scientific and technical activity where the risks are perceived to be substantial. These include the

---

[118]http://www.oecd.org/about/0,3347,en_2649_34141_1_1_1_1_1,00.html, visited 10 Sept 2009.

[119]http://www.oecd.org/document/7/0,3343,en_2649_34273_1911303_1_1_1_1,00.html, visited 10 Sept 2009.

[120]Public orchestration of private regulatory activity like that suggested here reflects the New Governance model of regulation. See Abbott & Snidal, Strengthening International Regulation, note 31, supra.

[121]Rappert, Chapter 8, this volume. Professional codes have begun to grapple with somewhat similar problems concerning the ethics of research.

chemical industry,[122] biological research – which many fear could be misused for destructive purposes such as biological weapons or terrorism[123] – and nanotechnology.[124]

In the latter field, for example, the Foresight Institute developed its Guidelines for Responsible Nanotechnology Development – applicable to scientific and technological researchers, business firms and even government policy-makers – over several iterations beginning in 1999, through workshops and other interactions among members of the nanotechnology development community.[125] The current Guidelines focus on the riskiest area of nanotechnology research, autonomous replicators, but also state broader principles of responsible development; separate guidelines apply to researchers, firms and policy-makers. In addition, the Nanotechnology Industries Association, in collaboration with the Royal Society and Insight Investment, and subsequently with the UK government-sponsored Nanotechnology Knowledge Transfer Network, adopted the Responsible NanoCode in 2008. The NanoCode consists of broad strategic principles that apply throughout the product life-cycle, supplemented by concrete examples of good practice in implementing each principle; good practice guidelines allow for the benchmarking of individual organizations' performance. The founding partners of the NanoCode are themselves widely representative, and the multi-stakeholder working group that developed the Code engaged in extensive consultations.[126] Public authorities such as the European Commission have also begun to promulgate voluntary nanotechnology codes.[127]

FASTIR could encourage the adoption of appropriate self-regulatory and other voluntary codes; facilitate their adoption and implementation; provide an international imprimatur for codes that meet agreed procedural and substantive standards;

---

[122]See, e.g., Responsible Care Global Charter, adopted by the International Council of Chemical Associations, http://www.responsiblecare.org/page.asp?p=6341&l=1, visited 10 Sept 2009.

[123]See The Royal Society. The role of codes of conduct in preventing the misuse of scientific research, RS policy document 03/05, available at http://royalsociety.org/document.asp?id=3215, visited 10 Sept 2009; Developments in codes of conduct since 2005, available at http://www.unog.ch/80256EDD006B8954/(httpAssets)/358F8EE5D00C281CC125747B004F57CF/$file/codes+background+paper+-+advanced+copy.pdf (biosecurity codes), visited 10 Sept 2009.

[124]See Bowman, Diana M., and Graeme A. Hodge. 2009. Counting on codes: An examination of transnational codes as regulatory governance mechanism for nanotechnology. *Regulation & Governance* 3: 145–64. In addition to codes adopted by firms or industry associations, multi-stakeholder groups and public authorities, the Coalition of Non-Governmental Organizations adopted "Principles for the Oversight of Nanotechnologies and Nanomaterials" in 2007. More aggressive than most of the other codes, e.g., in prescribing precautionary measures and calling for mandatory regulation, the Principles seek to shape the regulatory dialogue as well as the behavior of firms. Ibid.

[125]http://www.foresight.org/guidelines/current.html, visited 10 Sept 2009.

[126]http://www.responsiblenanocode.org, visited 10 Sept 2009; Information on the responsible nano code initiative, May 2008, available at id.

[127]See European Commission Adopts Code of Conduct for Responsible Nanosciences and Nanotechnologies Research, IP/08/193, Feb. 8, 2008, http://europa.eu/rapid/pressReleasesAction.do?reference=IP/08/193&format=HTML, visited 10 Sept 2009.

and disseminate information regarding the design and implementation of high-quality codes. The goal would be to create a "race to the top," in which industries, researchers and professions compete to be seen as appropriately regulating their activities.[128] The institutions of FASTIR, e.g., its secretariat and scientific advisory committee, could pursue this goal by working with existing groups. Those institutions could also participate in the process more directly, at least to some extent: for example, the convention's scientific and business advisory committees could develop model codes or work to to harmonize existing codes.

Trans-governmental networks perform a variety of regulatory functions, including facilitating enforcement. Most important for present purposes, "networks of agencies negotiate, implement, and diffuse norms that are often precise and elaborate, and may be politically powerful though not binding as a matter of [international law]. . . ."[129]

Many international financial regulatory regimes operate trans-governmentally. An important example is the Basel Committee on Banking Supervision, made up of central bank officials and banking regulators from major economies. The Basel Committee has formulated widely implemented guidelines for judging the adequacy of bank capital, as well as principles of effective bank supervision and norms allocating national jurisdiction to supervise international banks.[130] Trans-governmental norms such as the Basel guidelines are closer to "soft law" than to treaties or protocols, but they can be adopted more rapidly than treaties, are more flexible, and better reflect technical expertise. They can, moreover, be powerful mechanisms for learning, socialization and technical assistance.[131] To be sure, trans-governmental networks have significant weaknesses, especially when the participating agencies have conflicting interests or face domestic legal and political constraints.[132] Nonetheless, they would be a valuable element in an overall system of coordination, steering and facilitation. Under FASTIR, the same trans-governmental bodies that share information and formulate best practices in the use of scientific information and the design of regulatory processes could develop more specific substantive norms to steer and coordinate national regulations.

Finally, at the *international* level, this chapter has already discussed the framework convention-protocol approach to the incremental adoption of legally binding inter-state rules. Yet protocols need not be limited to the traditional inter-state

---

[128] Abbott & Snidal, Strengthening International Regulation, note 31 supra.

[129] Abbott, Kenneth W. 2008. Enriching rational choice institutionalism for the study of international law. University of Illinois Law Review 5: 27.

[130] Ibid., at 27. The 2008 financial crisis revealed the Basel capital adequacy guidelines to be *substantively* deficient. However, there is no reason to conclude that the *trans-governmental structure* of the Basel Committee was responsible for the deficiencies; indeed, the widespread adoption of the guidelines, which may have contributed to the breadth of the crisis, demonstrates the power of the trans-governmental approach.

[131] Slaughter, supra note 12, at 3–5, 36–55.

[132] See Verdier, Pierre-Hugues. 2009. Transnational regulatory networks and their limits. *Yale International Law Journal* 34: 113–172.

approach of treaties, in which implementation is left almost wholly to individual states. Instead, FASTIR protocols could themselves incorporate elements of the transnational and trans-governmental approaches.

9.8 Conclusion

Keeping pace with rapid scientific and technological innovation is one of the most challenging problems facing modern society. It is a multi-faceted problem, involving environmental, health and safety issues, economic issues, even social, cultural and ethical issues. It is fraught with uncertainty. Its structure poses difficult political and institutional challenges: it requires preliminary action before any concrete problems have appeared, then very fast action once problems do appear. Exacerbating all of these difficulties, it is a transnational problem, not merely a national one, and its transnational character is expanding as more and more countries develop capacities for innovation.

No institutional innovation will meet these challenges perfectly. But the framework convention approach proposed here would engage a wide range of governance approaches to address the many facets of the problem; would reduce uncertainty by promoting the sharing of information, common assessments and normative deliberations; and would facilitate institutional action by easing both initial preliminary responses and more concrete, adaptive regulatory responses over time. While imperfect, it may be the best transnational response to the "wicked" problem of innovation.

Chapter 10
Principles-Based Regulation and Emerging Technology

Ruth B. Carter and Gary E. Marchant

Innovations in emerging technologies are changing "every aspect of human life" at an unprecedented pace (NanoAction 2007). The companies trying to apply these advancements and the regulatory agencies trying to oversee these technologies are struggling to keep up the speed of development of new technologies, and need new approaches to be able to adapt to the complexity and rapid pace of innovation (U.S. Dep't of Treasury 2008). One possible mechanism to allow both innovators and regulators to remain more flexible and adaptable to rapidly changing technologies is to replace the current rules-based regulatory system with a principles-based approach. A principle-based approach has the potential to reward companies for acting ethically while encouraging innovativeness, efficiency, and competition in developing areas of technology. At the same time, a principles-based approach raises a number of difficult legal, policy and political challenges that would need to be overcome for such a system to succeed. This chapter will explain the background, objectives and experience of principles-based regulation, review the strengths and weaknesses of such an approach, and consider the applicability of principles-based regulation for the oversight of emerging technologies.

10.1 What Is Principles-Based Regulation?

There is a long-standing tension in regulatory theory between rules-based approaches and standards-based approaches to regulation (Kaplow 1992). Briefly, rules-based approaches set forth the specific acts or behaviors a regulated party is expected to achieve with some specificity (e.g., discharge no more than 0.45 pounds of pollutant X per hour) whereas a standards-based approach sets forth a more general goal that the regulated party must strive to achieve (e.g., discharge pollutant X at a level that will be safe for human health and the environment). Both approaches have their relative strengths and weaknesses – the key ones being that rules provide

R.B. Carter (✉)
Arizona State University, Tempe, AZ, USA
e-mail: ruthbcarter@gmail.com

G.E. Marchant et al. (eds.), *The Growing Gap Between Emerging Technologies and Legal-Ethical Oversight*, The International Library of Ethics, Law and Technology 7,
DOI 10.1007/978-94-007-1356-7_10, © Springer Science+Business Media B.V. 2011

more certainty but are also more rigid, whereas standards are more subjective and subject to different interpretations, but provide greater flexibility for adapting to shifting circumstances (Coglianese et al. 2003). In the legally-oriented regulatory system of the U.S. and increasingly other industrial nations, rule-based systems have been the preferred approach, often described as the "command and control approach." Under this approach, regulation is described as "the promulgation of rules by government accompanied by mechanisms for monitoring and enforcement" (Black 2002).

Principles-based regulation goes one step beyond standards and instead of specifying specific requirements (rules) or general obligations (standards), it focuses on desired outcomes (Barrass 2007; Black 2008). This approach uses guiding principles that are broad, general, and abstract (Cunningham 2007). Principles-based regulation is a "complex form of regulation" where expectations are expressed using qualitative terms and the underlying reason for them is given (Black 2008). Using principles, rather than rules, gives each company the discretion to determine how they can best apply them to their practices (Freshfields Bruckhaus Deringer 2007). As such, principles-based regulation is sometimes referred to as "light touch" oversight (Gray 2009).

As the principles are applied to new situations, they become more developed and transparent. (Schwarcz 2008). These principles not only provide companies greater flexibility to comply with their current expectations, they also provide greater agility for adapting to new situations and contexts that arise in rapidly changing areas of activity (Better Regulation Task Force 2003). By being outcome-focused, principles-based regulation can, encourage more collaborative approaches and focus on finding solutions to problems instead of being overburdened by attempts to stay in compliance with an inflexible rules-based system (Ford 2008; Hopper and Stainsby 2006).

Under a principles-based approach, the regulatory agency identifies the general principles that companies are expected to comply with, and then each company develops its own interpretation, framework, and best practices for adhering to the applicable principles (Kovacevich et al. 2008). In other words, companies are expected to be self-reflective in determining how the principles should be applied to their practices to ensure compliance (Black 2008). Principle-based regulation is most applicable where it would be unduly burdensome if not impossible to create specific rules for every possible scenario (Ford 2008). They should thus be used when "the regulator 'knows the result [he] is trying to achieve but does not know the means for achieving it, when circumstances are likely to change in ways that the [regulator] cannot predict, or when the [regulator] does not even know'" the specific result it desires (Ford 2008).

Because principles are usually worded very generally and thus may be subject to different interpretations, a key prerequisite of a principles-based approach is that the regulatory and regulated parties are well-intentioned and prepared to trust and cooperate with each other. The shift from rules-based to principles-based regulation requires a significant change in how companies interact with their regulators

(FSA 2007). Instead of the regulator unilaterally deciding if the company is following the rules, the regulator and the company must work together to "determine how the regulatory outcomes envisage in a principle are to be reached." (Barrass 2007). Principles-based regulation thus requires a "close engagement" between the regulator and the company that is more intimate than traditional inspection and supervision (Black 2008; FSA 2007). Conversely, the regulator should articulate the guiding principles in a "flexible and dynamic" manner and has the responsibility to clearly communicate the goals and outcomes companies are expected to achieve (Ford 2008; FSA 2007). Moreover, the regulator should direct its energy toward the most pressing problems rather than enforcing against technicalities or minor disagreements (Better Regulation Task Force 2003). The regulator is also expected to regulate consistently, which allows companies to be more innovative without violating the principles (Better Regulation Task Force 2003).

Although a principles-based system gives companies more flexibility than a rule-based approach, the regulator continues to settle disputes, take enforcement actions, and administer sanctions when principles are breached (FSA 2008; Ford 2008; U.S. Dept. of Treasury 2008). The regulator always has the responsibility for ensuring that companies are complying with the applicable principles (Cunningham 2007). The regulator distinguishes between companies who occasionally make mistakes from those with more serious problems, and direct their resources to the latter (Ford 2008).

Principles-based regulation is a relatively new concept that received much governmental and scholarly interest recently after being adopted to regulate financial institutions, most prominently the United Kingdom and Canada (Black 2008; Cunningham 2007). In the United Kingdom, financial regulation, conducted by the Financial Services Authority (FSA), is based on eleven broad principles, such as conducting business with integrity, observing proper standards of market conduct, treating customers fairly, protecting client's assets, and dealing with regulators in an open and cooperative way (FSA 2008). These broad principles are then supplemented with some specific rules, as well as some illustrative examples provided by FSA as guidance of good and bad practices (Hopper and Stainsby 2006). The FSA only brings enforcement action when one of the eleven principles is "clearly breached" (Hopper and Stainsby 2006). The U.K.'s. principles-based regulatory approach to financial regulation came under considerable criticism with the recent high-profile failure of Northern Rock Bank even though the institution seemed to be complying with the regulators' principles, and resulted in an enhanced reliance on back-up rules to reinforce the general principles (Gray 2009). Similarly, financial regulations in Canada are based on broad principles such as prohibiting "unfair practices," but this principles-based approach begins to resemble rules when it provides specific definitions, clarifications, and requirements (Cunningham 2007). Officials in the European Union and United States have expressed interest and support for adopting a similar principles-based approach for financial regulation in their jurisdictions, although the recent economic crisis has moved consideration of such changes to the back burner (Black 2008).

10.2 Benefits of Principles-Based Regulation

A number of potential benefits are attributed to principles-based regulation, although there has been a tendency to exaggerate some of these benefits. Notwithstanding some hyperbole, there are some important potential benefits associated with principles-based regulation.

10.2.1 Regulatory Objectives More Important than the Letter of the Law

The most significant benefit of principles-based regulation is that it puts the overall objective of regulation before the letter of the law, and thus may increase the likelihood that the regulatory objective will be achieved while reducing the risk of becoming embroiled in technicalities that are not critical to the overall goal of the regulation (Cunningham 2007). Principles-based regulation encourages more effective regulation by discouraging "loophole behavior and checklist style approaches" (Ford 2008). Under traditional rule-based regulation, companies have incentives to engage in "creative compliance" whereby they technically comply with the letter but not spirit of the regulation (Black 2008; Cunningham 2007; Ford 2008). Even the best-drafted rules can never prevent or anticipate all possible misconduct, and so there will inevitably be gaps between the wording and spirit of the rule that a company could exploit (FSA 2007). Additionally, regardless of how carefully a rule is written, it will always be "both under- and over-inclusive in relation to the original principle" (Kershaw 2005, p. 605). Finally, "the more precise the rules, the more complex they become ... the greater the potential for internal inconsistencies in their application, and the more uncertain their application becomes in any particular circumstance" (Black 2008, p. 438). In contrast, because the focus of principles-based regulation is on the overall objectives of the regulation rather than specific wording and detailed regulatory prescriptions, the focus is more clearly on the desired outcome. Complying with high-level principles, instead of being distracted by the minutiae of rigid rules, reduces misconduct, distraction, and misdirection by regulated entities (Hopper and Stainsby 2006).

10.2.2 Greater Flexibility for Companies

A principles-based regulatory system puts the responsibility on each company to determine the best way to apply the principles to their objectives to ensure that they are compliant (FSA 2008; Black 2008; Ford 2008). Each company has the flexibility to determine how each principle applies to their products, practices and businesses, which if approached in good faith should permit more creative, effective and efficient compliance (Freshfields Bruckhaus Deringer 2007; U.S. Dept. of Treasury 2008; Kershaw 2005). A principles-based approach gives companies more options for conducting business and accomplishing both their and the regulatory goals (FSA 2008; Black 2008). Not only will this allow companies to be more efficient and creative in their compliance, but the additional freedom should also make companies more cooperative and willing to comply with the regulatory objectives (Ford 2008; Black 2008).

10.2.3 More Dynamic and Adaptive Regulation

Another key advantage of principles-based regulation, of particular relevance to the topic of this book, is that principles-based regulation can better and more quickly adapt to changing circumstances than can traditional rules-based regulation (Black 2008). Because they are written in broad general terms, principles are usually resilient and maintain their relevance even as the regulated activities and their context change in response to evolving technologies and other factors (Cunningham 2007). In contrast, more detailed rules are often made obsolete or inapplicable by even relatively minor changes in circumstances. Rules are also unable to keep up with innovations and changes in the industry, and can easily become outdated (FSA 2007). Because rules require very specific and detailed language, supported by adequate legal and evidentiary analysis and process, rules take much longer to promulgate than principles and cannot be changed fast enough to effectively regulate industries that are constantly developing and changing (Cunningham 2007; Better Regulation Task Force 2003). The relative dynamism of principles-based regulation make it particularly well-suited for fast-changing regulatory subjects such as emerging technologies.

10.2.4 Better Relationship Between Regulators and Regulated Parties

The greater flexibility and reduction in detailed obligations provided by principles-based regulation may make companies more willing to accept regulation as "an integral part of business decision-making" (FSA 2008). Companies view the principles as something to internalize into their metrics of success instead of an obstacle to circumvent. Moreover, regulators are perceived more as an ally to accomplish mutual objectives together than an adversary to be feared and fought. Enforcement is cooperative, as the regulator considers reasonable efforts to apply the principles to each company's situation as substantial compliance, rather than bringing enforcement actions based on technical violations and paperwork discrepancies (FSA 2008; Barrass 2007). Multiple rules can be consolidated under one principle, making it easier for companies and regulators to understand what the expectations are and to ensure that compliance is occurring (Ford 2008). Furthermore, problems are resolved more efficiently because they are addressed more proactively and with input from the regulator (Kovacevich et al. 2008). From the perspective of regulators, the principles-based approach also provides greater flexibility as they only have to take enforcement action when a principle has been clearly violated (Black 2008).

10.3 Limitations of Principles-Based Regulation

Principles-based regulation is not a panacea, and has some important limitations, discussed next.

10.3.1 Uncertainty

One of the biggest drawbacks of principles-based regulation is the level of uncertainty associated with what is expected by applying the general principles to a wide variety of different situations (FSA 2008; Hopper and Stainsby 2007; Black 2008; Cunningham 2008). As they seek to apply the guiding principles to their own operations and activities, companies will be uncertain if their actions are in compliance with the regulator's understanding of the principles (Black 2008; Schwarcz 2008). The vagueness and flexibility of principles can lead to multiple interpretations in a particular context (Black 2008). Since the principles are open to interpretation, it will also be more challenging to identify when a principle has been breached (Barrass 2007), and there will be uncertainty how tolerant regulators will be of a company's divergent interpretation. One fear is that companies will be blamed in hindsight for actions that may have seemed like reasonable and good-faith interpretation of the principles at the time the decision was made (Gray 2009; Cunningham 2007). Alternatively, a regulator may feel constrained from taking enforcement action against a company that takes a self-serving approach to the principles that technically complies with the wording but not spirit of the principle (Gray 2009).

Using a principles-based system can lead to communication problems between the regulator and the regulated company (Black 2008). Companies will want to minimize their risk by requesting more direction and clear boundaries from the regulator (Black 2008). Yet, regulators will want to avoid undercutting the benefits of having broad principles by issuing too many interpretations and guidance on construing the principles. Alternatively, if the company receives too many communications from the regulator, they may be uncertain about what their expectations are (Black 2008).

10.3.2 Compliance Problems

Principles-based regulation relies heavily on companies being honest and open with their regulator (FSA Internal Audit Div. 2008). They are required to keep their regulator informed about changes and risks related to their endeavors. The regulator needs to provide sufficient supervision and conduct ongoing assessments on the companies' functions. In order to do this, the regulator requires more information than he would need in a rules-based system (FSA 2007). The regulator needs enough information and knowledge to fully understand the risks and choices each company is making (U.S. Dep't of Treasury 2008). Without such access and information, the government may not have enough knowledge about an industry, particularly when emerging technology is involved, to be able to identify when a company is not being compliant, the cause of the problem, or what solutions will resolve it (Black 2002). It is well-known that the "government cannot know as much about [an] industry as [the] industry does about itself." (Black 2002, p. 3). Hence, there may be a greater risk of non-compliance with principles than with a rule-based system.

A related compliance concern is that regulated companies must be motivated to perform beyond the minimal level of compliance to achieve the objectives of principles-based regulation (Black 2008). For at least some companies, this is likely to be a problem. Without detailed rules, it is harder for the regulator to convince recalcitrant companies to comply with regulations, and it is more difficult for the regulator to say "no" to companies' questionable practices (Black 2008).

In addition, the close relationship between companies and regulators necessary for principles-based regulation to thrive can lead to complacency or exploitation of the relationship. The regulator must be leery of companies that abuse the principles or their relationship with the regulator (Ford 2008). There is a risk that companies will attempt to use their relationship with the regulator to avoid punishments when they breach principles or commit fraud (Kovacevich et al. 2008). The closer relationship between the company and the regulator can result in the regulator becoming overly familiar with the company and can overlook breaches that he would have seen had he been more impartial (FSA Internal Audit Div. 2008). The principles-based system allows the regulator to provide less oversight to companies that have demonstrated compliance (Ford 2008). This can lead to an increase in abuse by companies. Adopting principles, rather than rules, results in fewer norms applying to an industry, which further opens the door for abuse by companies that can convince a regulator that their indiscretions are still in compliance with the guiding principles (Schwarcz 2008).

The bottom line is that a principles-based system can facilitate trust between the regulator and the company, but it cannot create it (Black 2008). "Compliance systems can be empowered under principles-based regulation, but only if they are strong already" (Black 2008, p. 427). If trust between the regulator and regulated does not already exist, a principles-based system can "never be operationalized." (Black 2008, p. 456).

10.3.3 Changes in the Industry's Culture

Changing to principles-based regulation requires a dramatic and perhaps risky shift in the culture of the industry (FSA 2007). Each company will shoulder more responsibility for how they meet their regulatory obligations (FSA 2007; Gray 2009). They will be "required to think through the application of the provisions to particular situations to a far greater degree than they are with respect to a detailed rule" (Black 2008, p. 440). There will be an increased need to exercise judgment by multiple levels of management to be in compliance with the guiding principles (FSA 2007). Companies will no longer be able to solely rely on their regulator for compliance information, but will have to make their own judgments regarding their ideas to create new technologies (FSA 2007). The pressure, freedom, and fear of setbacks due to compliance issues could lead companies to being overly cautious regarding the projects they take on (Schwarcz 2008; Cunningham 2007). This could potentially hamper innovation and stifle competition if companies are unwilling to take risks or are unwilling to do the research that will make their experiments and proposed products comply with the principles (Schwarcz 2008).

10.3.4 Principles Can Erode into Rules

There is a risk with principles-based regulation that, over time and through their application, the principles can begin to resemble rules (Gray 2009). Once the principles have begun to be applied, there can be a decrease in using them innovatively. Companies become conservative and only use the principles in limited ways that have been deemed to be in compliance in previous applications (Schwarcz 2008). They restrict their use of the principles rather than risk using them in more innovative ways that might be rejected by the regulator (Schwarcz 2008). If the regulator only accepts certain practices as complying with the principles or if companies treat the principles like detailed rules, the companies' application of the principles becomes homogenous (Black 2008). This canalization of principles into more rule-like requirements is especially likely when a regulator is highly punitive and inflexible in its enforcement policies (Black 2008). As the principles become more rule-like, the flexibility and innovation offered by a principles-based approach is lost.

10.4 Principles-Based Regulation for Emerging Technologies?

Although principles-based regulation has both significant benefits and limitations, a principles-based approach may have particular promise for regulating emerging technologies. Emerging technologies can arise and be deployed quickly, evolve and change at a rapid pace, and create a wide diversity of applications and contexts. Traditional rule-based regulation is hard-pressed to respond to all three of these challenges, whereas principles-based regulation provides the flexibility, speed and dexterity to deal with fast-moving and diverse regulatory situations. Some of the principles used in financial regulation could also apply to emerging technologies, such as conducting business with integrity and diligence (FSA 2007). Other relevant principles that have been used or proposed in other regulatory contexts include the need to promote innovation, increase efficiency, and enhance competition between companies (U.S. Dept. of Treasury 2008). Maintaining the safety of employees and consumers are also crucial principles for any company involved in emerging technologies (Kovacevich et al. 2008). Another potential principle that might apply to companies in emerging technology fields might be conducting business with a high degree of transparency (Better Regulation Task Force 2003).

A principles-based approach might be particularly apt as an interim regulatory approach for emerging technologies that can be instated relatively quickly and provide some oversight until a more traditional rule-based system can be developed. This can help to fill the oversight gap that often exists for emerging technologies while the government is developing the sufficient evidentiary base and knowledge needed to promulgate traditional rules (Marchant et al. 2008). For example, the European Union has adopted a code of practice for nanotechnology researchers that is based on general principles, and serves as a gap-filling measure until more concrete regulations can be adopted (European Commission 2008). Over time, the principles could be strengthened and "filled-out" into more of a rule-based system with increasing knowledge and experience (Better Regulation Task Force 2003). Nevertheless, a principles-based approach for emerging technologies would likely face many challenges, such as overcoming the culture of litigation that pervades the U.S. regulatory system (Black 2008).

10.5 Conclusion

Principles-based regulation for emerging technologies in the United States may help address the existing problem that rules-based regulation cannot keep up with the pace of new developments. Particularly if implemented as an interim approach while regulators develop more traditional rule-based approaches, principles-based regulation can serve in a flexible, adaptable, and dynamic gap-filling role. However, implementing a principles-based approach will not be without its challenges and problems. New technologies are being developed in various industries that are supervised by numerous, and sometimes overlapping, agencies. Applying a principles-based approach might require the creation of a separate agency that would be devoted solely to new technologies; which comes with the additional problem of determining when a technology or method is no longer considered "emerging." Otherwise numerous agencies will have to change to a principles-based approach to accommodate every industry where technology is advancing rapidly. Engineers who are developing new technologies will appreciate the freedom that comes with adhering to principles, but ultimately they may want the certainty that comes with having to adhering to rules.

References

Barrass, John. 2007. *A Matter of Principle*, Chartered Financial Analyst Magazine, Jan-Feb 2007, at 27.

Better Regulation Task Force. 2003. *Scientific research innovation with controls*. London: Better Regulation Task Force

Black, Julia. 2008. Forms and paradoxes of principles-based regulation. *Capital Markets Law Journal* 3: 425–457.

Black, Julia. 2002. Critical reflections on regulation. *Australian Journal of Legal Philosophy* 27: 1–35.

Coglianese, Cary, Nash Jennifer, and Olmstead Todd. 2003. Performance-based regulation: Prospects and limitations in health, safety, and environmental protection. *Administrative Law Review* 55: 705–729.

Cunningham, Lawrence A. 2007. A prescription to retire the rhetoric of "Principles-based systems" in corporate law, securities regulation, and accounting. *Vanderbilt Law Review* 60: 1411–1493.

European Commission. 2008. Commission recommendation of 07/02/2008 on a code of conduct for responsible nanosciences and nanotechnologies research, Brussels, C (2008) 424 final, available at http://ec.europa.eu/nanotechnology/pdf/nanocode-rec_pe0894c_en.pdf.

Financial Services Authority (FSA). 2007. *Principles-based regulation: Focusing on outcomes that matter*. London: FSA.

Financial Services Authority Internal Audit Div. 2008. *The supervision of northern rock: A lesson learned review*. London: FSA.

Ford, Christie L. 2008. New governance, compliance, and principles-based securities regulation. *American Business Law Journal* 45: 1–60.

Freshfields Bruckhaus Deringer. 2007. *FSA principles-based regulation*, available at www.freshfields.com/publications/pdfs/2007/17593.pdf.

Gray, Joanna. 2009. Is it time to highlight the limits of risk-based financial regulation? *Capital Markets Law Journal* 4(1): 50–62.

Hopper, Martyn, and Stainsby, Jenny. 2006. Principles-based regulation – Better regulation? Interantional Banking Law Review 7: 387–391.

Kaplow, Louis. 1992. Rules versus standards: An economic analysis. *Duke Law Journal* 42: 557–629.

Kershaw, David. 2005. Evading Enron: Taking principles too seriously in accounting regulation. *Modern Law Review* 68: 594–625.

Kovacevich, Richard M. et al. 2008. The financial services roundtable. *The blueprint for U.S. financial competitiveness: Executive summary*, available at www.fsround.org/cec/pdfs/FinalCompetivenessReport.pdf.

Marchant, Gary E., Sylvester, S. Douglas, and Abbott, Kenneth W. 2008. Risk management principles for nanotechnology. *NanoEthics* 2: 43–60.

NanoAction. 2007. Principles for the oversight of nanotechnologies and nanomaterials, available at www.nanoaction.org/doc/.

Schwarcz, Steven L. 2008. The 'Principles' paradox. Duke law school legal studies research paper series 205.

U.S. Dep't of the Treasury. 2008. The Department of the Treasury Blueprint for a modernized financial regulatory structure. Washington, D.C.: Government Printing Office.

Chapter 11
Administrative Law Tools for More Adaptive and Responsive Regulation

Lyn M. Gaudet and Gary E. Marchant

The "ossification" of regulatory rulemaking is a serious impediment for pacing regulation with rapidly changing technologies (McGarity 1992). As the regulatory process increasingly becomes bogged down by procedural requirements, evidentiary burdens, judicial review and other legal obligations, the gap between technology and regulation has continued to grow (Marchant, Chapter 2, this volume). The problem has two aspects. First, agencies are too slow to adopt regulations in the first place. Second, regulations that are in place quickly become out-dated and are not revised in a timely and efficient manner in response to changing technologies and circumstances. Obsolete regulations can be the equivalent to, or even worse than, no regulation at all. Accordingly, there are two dimensions of the improvements needed to make rulemaking better adapted to address rapidly emerging technologies: (i) regulate when needed in a timely manner and (ii) ensure the regulations that are in place remain current and address new issues or applications as they arise.

A number of innovative tools of administrative practice have been proposed or attempted to make the regulatory process more adaptive and synchronized with changing factual circumstances. This chapter will explore four administrative law tools that have the potential to make the regulatory process more responsive and adaptive to rapid technological change. The four tools addressed are (i) negotiated rulemaking, (ii) direct final rulemaking, (iii) temporary legislation and sunset provisions, and (iv) online or "e" rulemaking.[1] A brief history of each of the various techniques is provided, as well as some examples of the techniques' successes and failures. The discussion of these four tools is preceded by a general discussion of the rulemaking process and the problems it faces trying to stay current with rapidly changing technologies.

---

[1] This chapter will focus on the United States as a case study, but similar innovations in regulatory processes are being implemented or considered in other jurisdictions.

L.M. Gaudet (✉)
Arizona State University, Tempe, AZ, USA
e-mail: lyn.gaudet@gmail.com

G.E. Marchant et al. (eds.), *The Growing Gap Between Emerging Technologies and Legal-Ethical Oversight*, The International Library of Ethics, Law and Technology 7,
DOI 10.1007/978-94-007-1356-7_11,

11.1 The Challenge of Timely Rulemaking in a Dynamic World

Regrettably, some lessons can only be learned the hard way. The soundest proposal might be fraught with problems that will not or could not be realized until the proposal is actually implemented. Attempts at improving the rulemaking process is an area full of such lessons. It is an endeavor where innovative ideas based on logic and common sense do not necessarily translate into efficiency and success in real-world practice. One conclusion is clear from the history of trying to improve rulemaking – there is no magic bullet. While no one method is going to provide all the answers, there are nevertheless some potentially effective administrative tools that can be used to make regulatory rulemaking more adaptive and responsive to the challenges of rapidly emerging technologies. The techniques and methods discussed are for the most part not new – they have been tried and tested, in some cases with success and in others without – it is their application in the new context of oversight of rapidly emerging technologies that is the focus of the discussion that follows.

Regulatory agencies need to be able to adapt to changing times, new decisions, and changes in administrations (Kalen 2008). Emerging technologies, with their rapid pace of development, make it extremely difficult for agencies to keep current. To exacerbate this tension, while technology is changing the world faster than ever before, the regulatory process has, over time, become less rather than more efficient and consequently is unable to stay current with rapidly changing circumstances (Marchant, Chapter 2, this volume). In administrative law, executive and independent agencies create detailed regulations through rulemaking. A critical aspect of the regulatory process, rulemaking is unfortunately a time-consuming process and is one of the reasons it is difficult for agencies to respond promptly to change. The Administrative Procedure Act (APA), enacted in 1946, is the most important U.S. federal law that governs the administrative practice of U.S. regulatory agencies (Reigel and Owen 1982). The APA specifies the procedures federal agencies must follow when they issue rules or adjudicate cases. Under section 553 of the APA, a regulatory agency may engage in informal rulemaking by drafting and publishing a proposed rule that is then subject to public comment, a process described as "notice and comment" rulemaking (Reigel and Owen 1982).

Notice and comment rulemaking has become the predominant form of administrative policymaking in recent decades. While case-by-case adjudication was the dominant model of administrative practice in the initial decades under the APA in the United States, the advent of many new environmental, health and safety statutes and agencies in the 1970s required a more legislative informal rulemaking process that was better fitted to creating new, expansive regulatory programs (Stewart 1975; Stewart 2003). As this new era of social regulation emerged in the 1970s and 1980s, Congress continued to task regulatory agencies to address a broad swath of problems that affected or involved almost all private and federal activities, and regulation via informal rulemaking quickly became the standard approach to address these problems (Baram 1982).

Over time, however, it has become clear that traditional regulatory processes have become increasingly burdensome and inefficient. The most frequently cited drawbacks of the traditional APA rulemaking system were that it was slow, cumbersome, time-consuming, and resulted in too much litigation (Note 1981; Susskind and McMahon 1985). The notice-and-comment procedures encompass an adversarial process that has been referred to as the "regulate, litigate, regulate, litigate" syndrome (Coglianese 1997; Baram 1982; Holly-Walker 2007). As interested parties and their lawyers learned to strategically manipulate the rulemaking process and accompanying judicial review and regulatory analysis requirements, and as the technical and legal complexity underlying regulations increased, the regulatory process became increasingly slow and burdened. Over the past couple decades, Congress and the White House have added up to eighteen additional analytical requirements on agencies to consider the impacts of their regulations on entities such as small businesses, state and local governments, Indian tribes, environmental justice communities, and children, among others (Seidenfeld 2000; McGarity et al. 2010). While any given one of these requirements may seem reasonable, the cumulative effect has been to substantially burden and slow the rulemaking process. In addition, reviewing courts have frequently overturned or remanded rulemaking decisions on a variety of procedural or substantive grounds, creating what has been referred to as a "judicially created obstacle course" to regulation (Kalen 2008, p. 670).

These and other requirements resulted in the "ossification" of rulemaking, whereby promulgation of new regulations becomes increasingly delayed and difficult (McGarity 1992; Pierce 1995). As one commentator colorfully described this "sclerotic" regulatory process:

> The federal rulemaking process has become a lawyers' Elsysium, in which each regulatory proposal requires elaborate justification, generates voluminous comment, and requires in turn meticulous agency responses to every comment Any substantial change along the way requires a further comment period, and the full process often consumes years even prior to judicial review. Judicial review may add a further delay of years to the process of implementation .... If an agency error is found, the result is to remand the disputed regulation to the agency to start again, on the same glacial timetable. To describe this model is to mock it. It is a model that makes prompt regulatory action impossible; a model that dampens innovative approaches by the agency; a model that precludes timely correction or improvement of regulations once unfairness, mistakes, omissions, or better approaches are revealed; and a model that eliminates any vestige of the predictability or certainty that the regulated community seeks (Campbell 2008, 35).

It is these limitations and problems with the rulemaking process, including the length of time it takes to develop and promulgate regulations, that has spawned a variety of mechanisms to try to expedite and modernize rulemaking. While the outcomes of these various innovations are mixed, they do provide a set of tools that can and should be considered in trying to keep regulation current with rapidly developing technologies.

11.2 Negotiated Rulemaking

One of the most promising and innovative proposals to expedite and otherwise improve the rulemaking process was negotiated rulemaking or "reg neg" (Harter 1982). The underlying concept behind reg neg was simple but appealing – instead of having the agency unilaterally draft a regulation and then battle it out with interest groups after key decisions had already been made (even if officially only preliminarily), it would be more efficient and harmonious for the affected interests to sit down with the agency and negotiate the regulation from the outset. Unfortunately, the promise of reg neg has generally not been borne out by the empirical record of its implementation, although this procedure may remain a viable option in some limited circumstances.

Negotiated rulemaking entered the limelight in the late 1980s and early 1990s as a promising alternative to traditional rulemaking procedures for federal agencies to address the ossification problem (Harter 1982; Holly-Walker 2007). Negotiated rulemaking was conceived to promote collaborative bargaining among interested parties in order to formulate a proposed rule more quickly and harmoniously (Shuck 1979; Harter 1982). The process of drafting a rule through negotiation among the parties provided an alternative to the traditional "notice and comment" informal rulemaking, where a federal agency would formulate the proposed rule internally and only after the rule had already been drafted would the affected parties have an opportunity to give the agency their input (Harter 1982; Holly-Walker 2007). The underlying concept of regulatory negotiation was quite simple and appealing: it would be faster and more efficient to have the parties seek to reach a consensus up front and forego all the subsequent disputes and litigation that characterized traditional notice and comment rulemaking.

Negotiated rulemaking became the spearhead of a movement for regulatory reform in the 1980s (Baram 1982; Note 1981; Harter 1982). Almost one decade later, after only isolated attempts at regulatory negotiation, Congress adopted the Negotiated Rulemaking Act of 1990 (NRA 1990) that, while not requiring agencies to use regulatory negotiation, encouraged the use of reg neg by federal agencies and outlined the process for agencies that opted to use it. The NRA listed several important aspects of a planned regulation that makes it amenable to regulatory negotiation, including that: (a) "there are a limited number of identifiable interests that will be significantly affected by the rule;" (b) "there is a reasonable likelihood that a committee can be convened with a balanced representation of persons who – . . . can adequately represent the interests identified" above; (c) "there is a reasonable likelihood that a committee will reach a consensus on the proposed rule within a fixed period of time;" and (d) "the negotiated rulemaking procedure will not unreasonably delay the notice of proposed rulemaking and the issuance of the final rule" (NRA 1990, § 563(a)). In addition to the NRA, Congress has adopted more than a dozen subsequent statutes that require specific agencies to use negotiated rulemaking to create certain regulations. Affected agencies include the Departments of Education, Health and Human Services, Housing and Urban Development, and Interior, and the Nuclear Regulatory Commission (Coglianese 1997).

Despite predictions that negotiated rulemaking would be the solution to the problems of the traditional regulatory processes, it appears to have never reached wide-spread use among federal agencies and has turned out to play a very minor and not very successful role in the promulgation of federal regulation (Coglianese 1997). The process for negotiated rulemaking as laid out in the NRA and subsequent administrative practice is for an agency that has elected to utilize negotiated rule-making to announce that intention in a notice of proposed rulemaking (NPRM) published in the Federal Register, inviting organizations with an interest in participating in the reg neg to contact the agency if they wish to be a part of the negotiations. The agency will next determine a proposed list of participants for the negotiating committee and the interests they represent. The next step is selection of a neutral advisor, referred to as a convenor, who gathers the interested parties into the committee that will together negotiate the proposed rule. This convenor is usually a neutral party skilled in facilitation and resolution of multi-party disputes.

The goal of negotiations is to decide on a draft of the rule that all parties agree on. That agreement is then drafted by the agency into the text of a proposed rule. The proposed rule is published in the Federal Register, the traditional public notice and comment process is carried out, and the agency decides based on the comments received whether to modify the proposed rule according to public comment. It is worth noting that an agency that chooses to use negotiated rulemaking to draft a rule is still accountable for following all other APA procedures, developing a rule within its statutory authority, and explaining the result.

Unfortunately, actual experiences with regulatory negotiation did not live up to the high expectations placed upon it. Empirical assessment of negotiated rulemaking, and in particular a 1997 study by Professor Cary Coglianese, found that Reg Neg was rarely attempted and when undertaken neither saved time nor reduced litigation (Coglianese 1997). Over the 13 year period (1983–1996) evaluated by Coglianese, the overall proportion of agency regulations adopted using negotiated rulemaking was consistently small – less than one-tenth of one percent. The average negotiated rulemaking took approximately two and a half years to complete, the time measured was the time from which the agency announced its intent to form a negotiated rulemaking committee to the time the final rule was published, which if anything extended rather than shortened the time to promulgate a new regulation. Reg neg also did not seem to reduce rates of litigation against adopted regulations. Coglianese found that while the frequency of judicial challenge to environmental regulations adopted by traditional notice-and-comment rulemaking in 1987–1991 was approximately 26 percent, the litigation rates for negotiated rules increased rather than decreased as expected to somewhere in the range of thirty-five percent (Coglianese 1997). Other empirical studies likewise have found that reg neg generally slows down and delays, rather than speeding up, rulemaking, although it may produce some slight benefits in terms of the participants' perception of the rulemaking process (Langbein and Kerwin 2000; Freeman and Langbein 2000; Balla and Wright 2003; but see Harter 2000 for a contradictory perspective).

An example of the problems encountered in applying reg neg can be found in the 1990 Clean Air Act requirement that the U.S. Environmental Protection Agency (EPA) issue a rule mandating the use of oxygenated fuel to reduce urban smog in nonattainment areas, which EPA chose to implement using negotiated rulemaking procedures (Coglianese 1997). The EPA selected representatives from the automobile, petroleum, and renewable fuel industries, as well as from the environmental NGO community. After lengthy negotiations, the parties reached what one report described as a "nearly litigation-proof agreement" (Coglianese 1997, 1290). Yet within ten days of the rule's publication, the American Petroleum Institute (API) and Texaco, Inc. filed petitions for judicial review. Eventually an out-of-court agreement was reached and the EPA revised the rule. Shortly thereafter two other petroleum companies, both of which were represented (although indirectly) during the negotiations, challenged the rule. Again, an agreement was reached and the rule revised. Then the National Tank Truck Carriers (NTTC), a trade association representing approximately 200 common carrier fuel transporters, filed a petition for review against EPA. The NTTC had no representative, direct or indirect, during the negotiations. Yet again, the EPA and NTTC came to an agreement and the EPA agreed to revise the rule (Coglianese 1997).

Trouble for the reformulated gasoline rule continued when the API filed an administrative action against the EPA, which was ultimately unsuccessful, arguing the rule was inconsistent with the negotiated agreement and the Clean Air Act (Coglianese 1997). Lastly, the reformulated gasoline rule was the first U.S. regulation struck down by the World Trade Organization after Venezuela and Brazil successfully argued the foreign refiner baseline provisions in the reformulated gasoline rule enacted using reg neg were discriminatory and in violation of trade rules. Over three years after the rule's original adoption, the EPA was forced to revise it yet again. For a rule that was proclaimed to be a complete success story, it is clear that it was far from immune to controversy or helped to smooth the way for a faster, less contentious implementation (Coglianese 1997).

Another example is The No Child Left Behind Act of 2001 (NCLB), intended to "ensure that all children have a fair, equal and significant opportunity to obtain a high-quality education," and which mandated that the Department of Education (DOE) use negotiated rulemaking to formulate every rule promulgated under Title I of the Act (Holly-Walker 2007). The intended purpose of requiring negotiated rulemaking under the NCLB was to cultivate a relationship between state and federal governments, establish a consensus among the interested parties, and thereby improve the substance of the rules and improve the public education system (Holly-Walker 2007). Professor Holly-Walker has evaluated the performance of mandatory negotiated rulemaking under the NCLB, and has found that rather than accomplish any of these intended goals, negotiated rulemaking has been a hindrance, not a help, to the NCLB's implementation (Holley-Walker 2007).

A major flaw in the implementation of negotiated rulemaking under the NCLB has been the DOE's failure to create negotiated rulemaking committees that adequately represent the interests of parents and students (Holley-Walker 2007). For example, the statute required DOE to ensure an "equitable balance between representatives of parents and students and representatives of educators and education officials" to insure "that the views of both program beneficiaries and program providers are fairly heard and considered" (H.R. Rep. No. 107-334, at 809; Holly-Walker 2007). Yet, DOE failed to appoint to the negotiated rulemaking process any independent representatives for program beneficiaries, representing a significant failure in implementation of the Act (Holley-Walker 2007). Holley-Walker argues that exclusion of these groups was an intentional choice by the DOE to avoid the presence of parties whose views may conflict with the views that the DOE preferred to see promoted in negotiations. Such an intentional choice to exclude interested parties is in direct opposition to the negotiated rulemaking process (Holley-Walker 2007).

While negotiated rulemaking is theoretically sound, supported by valid intentions and intuition, the empirical implementation record of forming rules through negotiation is underwhelming, even downright disappointing. Forecast as the remedy for an ineffective and sluggish regulatory procedure, negotiated rulemaking seems to have fallen far short of its expectations. A vivid demonstration of the unpopularity of this tool is during the 2007 amendments to the Federal Food, Drug and Cosmetic Act, then Secretary of Health and Human Services Michael Leavitt wrote to Congress to oppose a proposed requirement for the FDA to use reg neg on the grounds that reg neg is too "time consuming and resource intensive" (Cited in Kobick 2010).

Notwithstanding the relatively dismal empirical record to date, it is conceivable that reg neg could be a potentially useful tool in limited situations (Note 1981). First, for a reg neg to work, all affected interests must be adequately represented in the negotiation process (Holley-Walker 2007). Regulatory issues that involve a large number of affected interests may therefore not be appropriate candidates for reg neg. Second, not only must all interested parties participate in the negotiations, they must be willing to negotiate in good faith, so some level of trust and spirit of engagement and cooperation is a necessary prerequisite for meaningful negotiation. Third, the subject of the reg neg must be ready and appropriate for negotiation, meaning that the subject matter must be sufficiently developed and narrow enough in scope that the parties can realistically resolve it. For at least some regulatory issues involving rapidly emerging technologies, these conditions may apply.

If these prerequisites for successful negotiation are met, regulatory negotiation may provide an opportunity for agencies to improve the slow, expensive, and ineffective traditional system of regulation in specific, appropriate contexts. Regulatory negotiation allows for informed debate, encourages parties to make concessions for the greater good, and provides a forum for stakeholders to advocate for what they consider to be important provisions, and most importantly, involves them in the decision making process. These aspects of negotiated rulemaking, when present, can improve the chances that a rule created through collaboration has a higher likelihood of acceptance by the parties that helped create it.

11.3 Direct Final Rulemaking

The second administrative tool to be discussed is direct final rulemaking (DFR). Pioneered by the U.S. Environmental Protection Agency in 1981, DFR was developed in the same time period time and for the same reasons (i.e., to make rulemaking more efficient) as negotiated rulemaking (Levin 1995). While negotiated rulemaking attempts to gather as much input as possible before proposing a rule and allow interested parties to participate in the drafting of the rule itself in order to reduce the likelihood of disagreement, DFR seeks to essentially bypass the entire front-end of the rulemaking process by having the agency draft the rule and push it through without any public input or comment until the rule is final. While the philosophy underlying the two tools is somewhat inapposite, the overall goal of both policies is the same – namely to promulgate rules as quickly as possible by seeking to minimize disruption and controversy.

DFR begins with the publication of a presumptively final rather than proposed rule (Noah 1999). The agency publishing the rule specifies a period of time, typically thirty days, in which it will accept comments or notice of comments (Noah 1999). If no comments are received within the thirty day period, then after another specified period of time (e.g., sixty days), the rule goes into effect and has the status and force of law (Noah 1999). Alternatively, if the agency receives an adverse comment or even notice of intent to file an adverse comment, the agency must withdraw the final rule and revert back to normal informal rulemaking and the publication of a proposed rule (which agencies often publish simultaneously with the direct final rule to hedge their bets) (Noah 1999). DFR thus allows an agency to dispense with some of the procedural aspects of rulemaking for rules that it expects to be uncontroversial, thereby speeding up the rulemaking process for non-controversial regulation and preserving staff resources and time for more disputed rulemakings (Kolber 2009).

Praised by Vice-President Al Gore and recommended by the Administrative Conference of the United States in the 1990s, DFR has been touted as an advantageous tool that regulatory agencies should utilize whenever feasible (Levin 1995; ACUS 1994). There is also recent evidence to suggest that Congress favors abbreviated rulemaking. In the Energy Independence and Security Act of 2007, Congress called for "accelerated" rulemakings and "direct final" rules as long as the agency does not receive opposition (Kalen 2008).

There are two possible legal justifications for DFR under the APA rulemaking provisions. The first is the good cause exemption of the APA, under which an agency may avoid using informal rulemaking procedures if doing so is impracticable, unnecessary, or contrary to the public interest (5 U.S.C. 553(b)(3)(B)). Under this rationale, if no member of the public desires to submit a comment, then providing an opportunity for such comment arguably meets the good cause exemption for "unnecessary" procedures. An alternative justification for DFR is that it may "substantially comply" with APA's informal rulemaking procedures and any technical non-conformance is harmless error (Levin 1995; Kolber 2009). Both legal defenses rely on the opportunity for an interested member of the public who disapproves of the substance of a proposed DFR to submit a negative comment, thereby forcing the agency to withdraw the rule and initiate traditional proceedings (Levin 1995). A direct final rule would therefore only be promulgated and finalized in cases where there were no negative comments, buttressing both the good cause and substantial compliance justifications. Nevertheless, neither of these two defenses has been judicially endorsed, and scholars have pointed to weaknesses of both approaches, which leaves the legality of DFR in question (Levin 1995; Kolber 2009).

The questionable legality of DFR has not stopped U.S. federal agencies from utilizing the procedure. DFR has been used most extensively by the EPA, which not only originated the concept, but also is the most prolific and longest user of DFR (Levin 1995). First used for revisions to state implementation plans (SIPs) under the Clean Air Act, which require frequent and often uncontroversial regulatory modifications, EPA experienced success with this first application of DFRs, having to withdraw fewer than five percent of 90 SIP revisions it proposed over a six-month period (Levin 1995; ACUS 1994). EPA also successfully used DFR to promulgate over 100 significant new use rules (SNURs) under the Toxic Substances Control Act (Levin 1995). In 1993 it was reported EPA experienced over a ninety percent success rate with DFR (Office of the Vice President 1993).

Initially, EPA's implementation of the DFR procedure was very straightforward, beginning with a publication of a direct final rule in the Federal Register, allowing thirty days for comment and if no comments or notice of comments were received, the rule became law after sixty days (Levin 1995). In the mid-1990s, the EPA began to tweak its DFR procedure. The EPA began publishing two notices of proposed rulemaking: one of its intent to adopt a DFR and one of traditional informal rulemaking. In the event adverse comments are received in response to the DFR, the parallel notice of proposed rulemaking allows the agency to proceed uninterrupted with the rulemaking, but it must then follow the traditional notice-and-comment rulemaking protocol.

EPA has also been creative in expanding the use of DFR to cover some pieces (but not others) of major regulatory actions (Levin 1995). An example of such an undertaking was when EPA made significant revisions to its major reformulated gasoline program (RFG) mandated by the 1990 Clean Air Act Amendments just five months after it adopted the original regulation. The EPA argued the DFR made only minor modifications to the RFG rule, but when the agency received adverse comments, instead of withdrawing the DFR as a whole it withdrew only the portions of the rule to which the adverse comments were directed (59 Fed. Reg. 36,944, 1994; see also Levin 1995). Another example of EPA using DFR in a piecemeal way was regarding written exemptions from the acid rain program (Levin 1995). The DFR provided exemptions from the permitting and monitoring requirements for twenty-six plants. The agency treated the exemptions for the individual plants as severable, explaining it would only withdraw exemptions that received adverse comments, allowing the others to stand (60 Fed. Reg. 4413, 1995).

Other agencies have used DFRs, but with less frequency and success than EPA. For example, 40 percent of DFRs proposed by the FDA have had to be withdrawn due to significant opposition (Kolber 2009). It has been suggested that this high rate of unsuccessful deployment of DFR is due to the FDA's poor prediction of which rules will not be controversial and its use of the procedure in ways not intended (Kolber 2009). The FDA's poor record with DFR has led one critic to raise "real concerns about the value and wisdom of the innovation" (i.e., DFR) (Kolber 2009). Between the years 1993 and 1993, a total of 1,030 DFRs were proposed by federal agencies with only 62 published rules subsequently being withdrawn or removed in whole or part (Noah 1999). Such statistics suggest that the FDA may be an anomaly with its poor track record with the procedure, lending credence to the argument that DFR can be used effectively if applied in the appropriate circumstances to proper material.

In summary, there is some disagreement among scholars as to whether DFR is a valuable and effective rulemaking tool (Levin 1995; Noah 1999; Kolber 2009). The most important factor in predicting whether DFR will or will not be successful seems to be whether the subject matter is appropriate for this type of procedure as well as being non-controversial. DFR seems best suited for minor changes in language or the adoption of Congressional mandates and may not be appropriate for entirely new rules. While the jury is still out on the ultimate utility of DFR, it can be a useful tool for expediting agency rulemaking in certain and limited circumstances. If DFR is going to be used more widely in the future, it has been recommended that federal agencies and the Office of Management and Budget produce some guidelines as to the procedure's intended uses and best practices (Kolber 2009).

11.4 e-Rulemaking

The rise of the internet, social media, and other forms of communication technologies are creating new opportunities for more responsive, dynamic and speedy regulation. According to one leading expert, "informal rulemaking ... is about to be transformed by the silent revolution of e-government, the widespread incorporation of Web-based technology in the public sector" (Noveck 2004). "e-Rulemaking" is simply the use of digital technologies to develop and implement regulations (Coglianese 2004). More broadly, it includes the use of information technologies to facilitate a number of activities related to the process of developing regulations, including expanding public comment and participation in the rulemaking process (Copeland 2008). The primary benefits of incorporating digital technology, specifically the internet, in the rulemaking process are twofold: first, access to a large quantity of information from a large number and variety of sources, the sheer volume and diversity of which would not be possible through traditional methods, and second, increased opportunities for public participation in the rulemaking process (Coglianese 2004; Noveck and Johnson 2008; Johnson 1998).

In the words of the American Bar Association's Committee on the Status and Future of Federal eRulemaking, "new information and communication technologies could be applied in federal agency rulemaking to enhance public participation, make the process itself more efficient for both the public and the government, and ultimately produce better decisions" (ABA 2010). Moreover, the new capabilities of online tools can also be used to increase agency transparency and accountability, which can help build trust and inspire public confidence (ABA 2010).

Finally, and most relevant for the present purpose, e-Rulemaking has the potential to expedite rulemaking processes and outcomes. Digital technologies "may help streamline and improve regulatory management, such as by helping agency staff retrieve and analyze vast quantities of information from diverse sources"

(Coglianese 2004, 355). Indeed, one of the metrics proposed to measure the effectiveness of e-rulemaking is the amount of time it takes to develop a rule (Coglianese 2004).

E-rulemaking has been actively pursued in both the United States and the European Union over the past decade, with initiatives progressively getting more innovative and ambitious over time. It was first introduced in the U.S. on an individual agency basis, with the EPA and Department of Transportation (DOT) being the first to provide an opportunity for electronic submission of rulemaking comments. In 2002, Congress enacted the E-Government Act that required all federal agencies to accept public comments electronically and create one or more federal websites where the public can access those comments and other materials relevant to pending rulemakings (E-Government Act of 2002). There are two phases to the E-Government initiative, the first was the creation of a website in 2003 where federal rules currently open for comment could be located and comments made (www.regulation.gov), while the second phase, designed to allow the public to see other agency rulemaking materials, such as cost-and-benefit analyses for the rule, is currently being implemented (Sunstein 2010).

Most commentators believe the implementation of e-rulemaking in the United States has been limited and perhaps even disappointing to date (Noveck 2004; Benjamin 2006; Noveck and Johnson 2008; ABA 2010). To date, e-Rulemaking initiatives primarily consist of just accepting citizen comments at an online website. As currently implemented, e-rulemaking may actually slow rather than expedite rulemaking to the extent that the more convenient mechanism for submitting public comments will deluge the agency with more comments, much of them of a junk variety (Benjamin 2006; Noveck 2004; Noveck and Johnson 2008). But, e-rulemaking holds enormous promise for more creative, collaborative rulemaking approaches that may enhance and expedite regulatory decision-making (Coglianese 2004). Possibilities include online regulatory negotiations or juries, digital public hearings, improved data mining capabilities, integrative tools, and many other proposals (Coglianses 2004; Noveck 2004; Noveck and Johnson 2008). Some recent initiatives from the White House have been pushing greater emphasis and role for electronic rulemaking (Sunstein 2010). So, while e-rulemaking has yet to make a significant improvement in expediting regulation, it does hold significant promise for creative application to improve and accelerate rulemaking, and is thus a potentially valuable addition to the administrative rulemaking toolbox.

The European Union has also undertaken e-rulemaking as part of its Interactive Policy Making (IPM) initiative, which seeks to use the internet and other new technologies to improve communication between the EU, its member state governments, stakeholders, and the general public. The goals of the IPM initiative are to "assist policy development by allowing more rapid and targeted responses to emerging issues and problems, improving the assessment of the impact of policies (or the absence of them) and providing greater accountability to citizens." (European Commission 2010a). Started in 2001, the IPM initiative has evolved over time, with one of the more recent developments being the creation of a "single access point" called "Your Voice in Europe" that serves as a centralized site for all consultations, discussions and other tools in which the public and stakeholder scan play a role using the internet in the EU's governance (European Commission 2010b). As in the United States, the EU initiative has consisted of a relatively simple one-way communication channel by which citizens can provide comments on government initiatives over the internet, although the potential for more creative and innovative forms of electronic interaction are being explored.

11.5 Temporary and Sunset Legislation

Temporary legislation may be one of the most overlooked and underutilized administrative methods for addressing rapidly changing fields or technologies (Gersen 2007). Temporary legislation is particularly apt for regulatory areas replete with uncertainty, which certainly applies to most emerging technologies. Flexibility is critical when dealing with the unknown. The key advantages temporary legislation provides are the opportunity for a quantity of information to be incorporated into legislation and allowance for initial experimentation and a subsequent adjustment of policies. Consequently, the appeal of temporary legislation is greatest in areas of newly recognized risks (Gersen 2007).

Temporary legislation is characterized as laws that will expire at a specific date, referred to as the sunset date, without affirmative legislative action. Designed to be an active tool, this natural expiration date is intended to force the legislature to revisit the issues and assumptions contained in the legislation to determine whether it is worthy of renewal as well as to make necessary revisions. For example, when the National Bioethics Advisory Commission proposed a ban on cloning humans, it recommended that "[i]t is critical, however, that such legislation include a sunset clause to ensure that Congress will review this issue after a specified period of time (three to five years) to decide whether the prohibition continues to be needed" (Shapiro 1997).

Unfortunately, the empirical record suggests that Congress has often complied with the technical requirements but not the spirit of temporary legislation. If Congress so wishes, it can decline to meaningfully engage in reexamination of legislation and simply allow it to continue by giving it an empty blessing. An example is the Clean Water Act, which is required to be reauthorized every five years but it receives a rubber stamp approval every time it would otherwise expire because Congress is reluctant to address the significant policy issues and shortcomings of the current statute. To be truly effective, temporary legislation needs a hammer, a provision that forces regulators to deal with the substance of the regulation.

Temporary legislation and sunset clauses are included in a large number and diverse areas of subject matter, including environmental law, internet law, gun control, the PATRIOT act, and tax law. Sunset provisions have also been applied with some success in international agreements, such as the world trade laws. In 1947 the General Agreement on Tariffs and Trade (GATT) was formed by twenty-two countries, the United States among them (Gutterman and Brown 2009). Between 1986 and 1994, the "Uruguay Round" took place and included negotiations on trade in goods and services in a number of areas, including agriculture, intellectual property rights and counterfeit goods, textiles, investment policies and dispute resolution. The results of the Uruguay Round included the establishment of a series of Multilateral Trade Agreements (MTAs). These MTAs are binding on the members of WTO and address a number and breadth of areas, some of which are subsidies and countervailing measures, sanitary and phyto-sanitary measures, technical barriers to trade, anti-dumping measures, trade-related investment measures, and safeguards. The Safeguards Agreement is where sunset clauses come into play. A WTO member is allowed to implement a "safeguard" action, an example is a temporary restriction on a particular product, in order to protect a domestic industry from serious injury from an unforeseen increase of imports. The WTO recognizes the need for a country to be able to provide itself such protection but recognized that these safeguards could potentially lead to unfair trade practices and so the Safeguards Agreement has some built in safeguards of its own in the form of limitations and sunset clauses. Every safeguard action has a built-in sunset clause as well as the general prohibition of any grey area safeguard actions. The Sunset Agreement allows a country to implement a protective safeguard but the safeguard must be specific, targeted to a specific product, and the safeguard will automatically terminate after a certain amount of time. The combination of protection and limitation strikes a balance that addresses the needs and concerns of both sides (Gutterman and Brown 2009).

A more limited form of temporary legislation is mandatory periodic review requirements. This type of provision requires the legislature, or the regulatory agency implementing the legislation, to conduct a review every two years (or some other specified period) to evaluate the progress and problems encountered under the legislation. An example is that the U.S. Clean Air Act requires the EPA to review each of the national ambient air quality standards every five years to ensure they reflect the most up-to-date scientific knowledge (Blais and Wagner 2008). Another example is California's zero emission vehicle mandate adopted in 1990, which required a biennial review process to ensure that the requirements were technologically and economically feasible (NRC 2006). While such a provision can provide a useful vehicle for midcourse correction, it can also destabilize regulatory programs and undermine certainty in the program due to the periodic re-examination and potential revision.

Indeed, the primary criticism of temporary legislation is that it undermines confidence and predictability in the regulatory scheme because the legislation is open to regular revision. But confidence in legislation simply because it is permanent may be pointless if that legislation is based on out-dated assumptions or facts. It is unrealistic for the public or for the legislature to have the expectation that they would get everything right the first time. Hindsight – not foresight – is twenty-twenty. Temporary legislation and sunset clauses are tools that can be used in certain areas when some action is needed, but there is a question as to what that something should be. Additionally, the natural expiration and affirmative renewal required by temporary legislation protect against erroneous beliefs and predictions about the future of a rapidly evolving area or emerging technology from being codified into permanent law.

11.6 Conclusion

This chapter has identified and discussed a number of administrative tools that have the potential to make rulemaking more expeditious or dynamic. Each of the administrative tools discussed in this chapter highlight one aspect of the rulemaking process and modifies it in some way. Despite these differences each of the distinct tools share a common goal – to improve a specific aspect of the overall regulatory process. The empirical record for each tool is more complicated and problematic than the theoretical case for the tool might have anticipated. Nonetheless, the real-life experience for each of the tools allows for a more realistic and nuanced view of the potential benefits and applications of each tool in appropriate cases.

References

American Bar Association, Committee on the Status and Future of Federal eRulemaking. 2010. Achieving the potential: The future of federal E-rulemaking. *Administrative Law Review* 62: 279–288.

American Bar Association Section of Administrative Law and Regulatory Practice. 2009. Improving the administrative process: A report to the president-elect of the United States (2008). *Administrative Law Review* 61: 235–247.

American Conference of the United States. 1994. Recommendation 93-4, "Improving the environment for agency rulemaking," *Federal Register* 59: 4670, 1 Feb 1994.

Balla, Steven, and Wright, John. 2003. Consensual rulemaking and the time it takes to develop rules. In *Politics, policy, and organizations*, eds. Kenneth Meier and George Krause, 187–206. Ann Arbor, MI: University of Michigan Press.

Baram, Michael S. 1982. *Alternatives to regulation*. Lexington, MA: Lexington Books.

Benjamin, Stuart Minor. 2006. Evaluating E-rulemaking: Public participation and political institutions. *Duke Law Journal* 55: 893–941.

Blais, Lynn E., and Wagner, Wendy E. 2008. Emerging science, adaptive regulation, and the problem of rulemaking ruts. *Texas Law Review* 86: 1701–1739.

Campbell, Bradley M. 2008. Landmarks and land mines. *The Environmental Forum*, Nov/Dec: 30–35.

Clean Water Act. 33 U.S.C. §1251 et seq. (1972).

Coglianese, Cary. 2004. E-Rulemaking: Information technology and the regulatory process. *Admin. L. Rev*. 56: 353–402.

Coglianese, Cary. 1997. Assessing consensus: The promise and performance of negotiated rulemaking. *Duke Law Journal* 44: 1255–1349.

Copeland, Curtis. 2008. Congressional Research Service Report RL32240 The Federal Rulemaking Process: An Overview.

E-Government Act of 2002, section 206(c) & (d)(1)-(2). Pub. L. No. 107–347, 116 Stat. 2899.

European Commission. 2010a. What is IPM (Interactive policy making)?, available at http://ec.europa.eu/yourvoice/ipm/index_en.htm.

European Commission. 2010b. Your voice in Europe, available at http://ec.europa.eu/yourvoice/index_en.htm.

Freeman, Jody and Langbein, Laura. 2000. Regulatory negotiation and the legitimacy benefit. *NYU Environmental Law Journal* 9: 60–151.

Gersen, Jacob E. 2007. Temporary legislation. *University of Chicago Law Review* 74: 247–297.

Gutterman, Alan S. and Brown, Robert L. 2009. *Going global: A guide to building an international business*. Thomson West.

Harter, Philip J. 1982. Negotiated regulations: A cure for malaise. *Georgetown Law Journal* 71: 1–118.

Harter, Philip J. 2000. Assessing the assessors: The actual performance of negotiated rulemaking. *New York University Environmental Law Journal* 9: 32–59.

Holley-Walker, Danielle. 2007. Importance of negotiated rulemaking to the No Child Left Behind Act. *Nebraska Law-Review* 85: 1015–1057.

Jonhson, S. M. 1998. The internet changes everything: Revolutionizing public participation and access to government information through the internet. *Administrative Law Review* 50: 277–337.

Kalen, S. 2008. The transformation of modern administrative law: Changing administrations and environmental guidance documents. *Ecology Law Quarterly* 35: 657–720.

Kobick, Julia. 2010. Negotiated rulemaking: The next step in regulatory innovation at the Food and Drug Administration? *Food and Drug Law Journal* 65: 425–445.

Kolber, Michael. 2009. Rulemaking without rules: An empirical study of direct final rulemaking. *Albany Law Review* 72: 79–113.

Langbein, Laura and Kerwin, Cornelius. 2000. Regulatory negotiation versus conventional rule making: Claims, counterclaims, and empirical evidence. *Journal of Public Administration Research and Theory* 10: 599–632.

Levin, Ronald M. 1995. Direct final rulemaking. *George Washington Law* 64: 1–34.

McGarity, Thomas O. 1992. Some thoughts on deossifying the rulemaking process. *Duke Law Journal* 41: 1385–1462.

McGarity, Thomas, Steinzor Rena, Shapiro Sidney, and Shudtz Matthew. 2010. Workers at risk: Regulatory dysfunction at OSHA, Center for progressive reform white paper #1003.

National Research Council. 2006. *State and federal standards for mobile-source emissions.* Washington, DC: National Academy Press.

Negotiated Rulemaking Act (NRA). 1990. Pub. L. No. 101–648, 104 Stat. 4969, codified at 5 U.S.C.A. § 561.

Noah, Lars. 1999. Doubts about direct final rulemaking. *Admininistrative Law Review* 51: 402–428.

Note (1981). Rethinking regulation: Negotiation as an alternative to traditional rule making. *Harvard Law Review* 94: 1871–1891.

Noveck, Beth S. 2004. The electronic revolution in rulemaking. *Emory Law Journal* 53: 433–522.

Noveck, Beth S., and Johnson, David R. 2008. A complex(ity) strategy for breaking the logjam. *New York University Law Review* 17: 170–193.

Office of the Vice President. 1993. Accompanying report of the national performance review: Improving regulatory systems.

Pierce, Richard J. Jr. (1995). Seven ways to deossify agency rulemaking. *Administrative Law Review* 47: 59–95.

Negotiated Rulemaking Act of 1990. Pub. L. No. 101-648, 104 Stat. 4969 (1990).

Administrative Dispute Resolution Act of 1996. Pub. L. No. 104-320, §§ 1, 11, 110 Stat. 3870, 3873–3874.

Reigel, S.A., and P.J. Owen. 1982. *Administrative law: The law of government agencies*. Ann Arbor, MI: Ann Arbor Science.

Seidenfeld, Mark A. 2000. Table of requirements for federal administrative rulemaking. *Florida State University Law Review* 27: 533–536.

Shapiro, Harold T. 1997. Ethical and policy issues of human cloning. *Science* 277: 195–196.

Shuck, Peter. 1979. Litigation, bargaining, and regulation. *Regulation* 3(4): 26–34.

Stewart, Richard B. 2003. Administrative law in the twenty-first century. *New York University Law Review* 78: 437–460.

Stewart, Richard B. 1975. The reformation of American administrative law. *Harvard Law Review* 88: 1669–1813.

Sunstein, Cass R. 2010. Memorandum for the president's management council: Increasing openness in the rulemaking process – Improving electronic dockets. Office of Management and Budget, 28 May 2010.

Susskind, L., and McMahon, G. 1985. The theory and practice of negotiated rulemaking. *Yale Journal on Regulation* 3: 133–165.

U.S. Dept. of Education. http://www.ed.gov/policy/highered/reg/hearulemaking/hea08/neg-reg-faq.html (last visited 5 June 2009).

World Bank Group. 2008. The World Bank, development and climate change – A strategic framework for the World Bank Group consultation draft 11. http://siteresources.worldbank.org/EXTCC/Resources/407863-1219339233881/DevelopmentandClimateChange.pdf

Chapter 12
Collaborative Voluntary Programs: Lessons from Environmental Law

Kathleen Waugh and Gary E. Marchant

Scholars have likened the technology revolution that we are currently experiencing to the Industrial Revolution that occurred in the early nineteenth century (Hirsch 2006; Isenberg 1995; Litan 2001). Both "revolutions" introduced new technologies that transformed society, providing almost unimaginable benefits; but accompanying the benefits were unfortunate side effects and consequences (Hirsch 2006; Pearson 2002).

In the case of the Industrial Revolution, the new technologies generated considerable harm to the environment, and a new form of law – environmental law – developed in response to that challenge. In the case of today's converging technological revolutions, the legal system is grappling with how to deal with new challenges created by the rapid advances in science and technology. Because environmental law frequently addresses the output or effects of technology, particularly as it relates to the harm caused society, the variety of new regulatory approaches that have been tried in the area of environmental law in the past 20 years may prove instructive as the legal system grapples with today's technology challenges. In particular, the environmental field has utilized a variety of innovative cooperative and voluntary programs to enhance or supplement the environmental benefits obtained through traditional regulation (Gunningham 2009b). This experience with voluntary programs provides a rich history to consider and evaluate potential voluntary approaches to the oversight of emerging technologies.

This chapter begins with a general background on voluntary and collaborative programs in environmental law, including different types and common elements of such programs. The next section summarizes the strengths and limitations of these voluntary approaches. The third section then reviews the empirical experience of several specific voluntary programs and the lessons that can be drawn from those examples for the governance of emerging technologies.

K. Waugh (✉)
Arizona State University, Tempe, AZ, USA
e-mail: Kathleen.Waugh@asu.edu

G.E. Marchant et al. (eds.), *The Growing Gap Between Emerging Technologies and Legal-Ethical Oversight*, The International Library of Ethics, Law and Technology 7,
DOI 10.1007/978-94-007-1356-7_12, © Springer Science+Business Media B.V. 2011

12.1 Background on Voluntary Environmental Programs

The traditional form of environmental regulation known as "command and control," in which the government adopts rules of performance that are then enforced against regulated parties, has increasingly been criticized as an overly rigid and cumbersome system that discourages technical innovation because of the focus on narrowly-defined compliance to uniform minimum standards (Wyeth 2006). In the 1980s, environmental law began to utilize voluntary collaborative programs as a means to address the criticisms of traditional regulation, including the problem of outdated rules and disincentives for innovation (Gunningham 2009a). In the United States, many collaborative voluntary programs developed under the auspices of the U.S. Environmental Protection Agency (EPA) and the Occupational Safety and Health Agency (OSHA). As of July 2009, EPA was sponsoring approximately 60 collaborative partnership programs with 13,000 participants, including firms, industry groups and other organizations (U.S. EPA 2009). OSHA had 2,245 facilities participating in the agency's voluntary protection programs, and there were 616 partnerships between OSHA with associations, industry groups, and businesses (OSHA 2007).

The utilization of collaborative voluntary agreements is not unique to the United States. In the European Union, 300 voluntary agreements were in existence as of the mid-1990s, and the number continues to grow (Johnson 2001). For example, several international initiatives have developed in response to the health, safety and environmental issues created by the emergent field of nanotechnology. International initiatives to address this uncertainty include the European Union's voluntary *Code of Conduct for Responsible Nanosciences and Nanotechnologies Research* and the development of voluntary standards by ASTM International and the International Organization for Standardization (ISO), among others (Breggin and Carothers 2006). The European Union's *Code* sets forth guidelines for conducting research that embrace the precautionary principle, emphasizing that research should be conducted in a safe and ethical manner while fostering the creativity and flexibility necessary to promote innovation and growth (CEC 2008). ASTM International and the ISO have undertaken the development of voluntary standards for characterizing the physical properties of nanomaterials and assessing the risks and environmental impact of the toxicological properties of nanomaterials (ASTM undated).

These collaborative and voluntary approaches in both the U.S. and E.U. are of several different kinds (Alberini and Segerson 2002). One type of collaborative model is "industry self-regulation", in which businesses voluntarily police themselves through "business-led initiatives" without regulatory intervention by the government. Another form is a voluntary government program in which the regulator determines the terms of the agreement, designs the program and eligibility requirements, and then seeks participants. A third model is a negotiated agreement between the regulator and the regulated entity, in which some form of incentive or regulatory relief (e.g., relaxing of permitting or inspection rules) is offered by the regulator.

While there are many differences between the various collaborative approaches, there are also many common characteristics. These characteristics include the following:

- Leadership comes from multiple sources – the government, the individual business, the industry sector, community groups, and environmental groups, although the government agency may still play the largest role (Wyeth 2006; Karkkainen 2006). This is unlike the traditional command and control model, where leadership is highly centralized in the rulemaking agency (Fiorino 1996).
- The process of working together to negotiate a solution that considers the needs of both society and that of an individual company or sector is seen as fostering more creative solutions than is typical of the top-down, adversarial approach of traditional regulation (Caldart and Ashford 1999). There is an underlying premise that a uniform, "one-size fits all" approach to regulation is not optimal (Hirsch 2001a).
- The process of negotiating a collaborative agreement shifts the focus from compliance to looking at the potential for continuous improvement through innovation (Wyeth 2006).
- Significant flexibility is offered in how a regulated entity meets performance objectives, and the programs delineate performance goals, not the technology to be used (Davies and Mazurek 1996). This flexibility is critical for fostering innovation.

Voluntary collaborative agreements are potentially useful in expediting an oversight mechanism for new problems as well as minimizing outdated regulations. More informed decisions often result from the collaboration, because the companies or industry usually know their processes and operations better than the government can, and the voluntary programs are often structured to require or encourage companies to disclose relevant information to regulators and concerned stakeholders (Wyeth 2006; Sousa and Klyza 2007). Moreover, an approach that allows entities the flexibility to determine how best to meet performance targets stimulates innovation because it removes the incentive to remain stagnant by simply maintaining compliance with static and often outdated standards (Gunningham 2009a; Hirsch 2006). In essence, such a system "self-corrects" over time. This freedom to innovate can be crucial to industries undergoing rapid technological change.

Self-regulation through voluntary collaborative programs has also generated concerns, however. The propriety of a federal agency negotiating standards with the regulated entity has been questioned, due to the risk that the agency may fall captive to special interest groups and thus compromise its "watch dog" mission and role as trustee of societal resources (Zinn 2002). Additionally, some collaborative programs have been criticized as being just as, if not more, bureaucratic and administratively burdensome as the traditional regulatory process, in which rulemaking may take several years (Hirsch 2001b; Davies and Mazurek 1996). Some collaborative programs have also floundered due to legal problems relating to doubts about the agency's statutory authority to enter into collaborative or other innovative programs

(Caballero 1998). There are lessons to be learned from the experiments with voluntary and collaborative programs in environmental law, and below we explore some of those programs and the lessons that might be drawn from them for emerging technologies.

12.2 Examples of Voluntary Environmental Programs

In this section, the design and results of a number and variety of voluntary or cooperative environmental programs will be summarized, with the goal of drawing some lessons that could be useful for the governance of emerging technologies.

12.2.1 33/50 Program

The EPA first entered the arena of voluntary programs with its "33/50" program, which was launched in 1991 (Kerret and Tal 2005). This was a voluntary government program in which the government determined the terms of the agreement, designed the program and eligibility requirements, and then solicited participants (Coglianese and Nash 2008). EPA approached approximately 8,100 businesses that emitted any of 17 hazardous air pollutants identified as a priority by EPA based on reported emissions under the Toxic Release Inventory (TRI) (Coglianese and Nash 2008). Businesses were asked to voluntarily reduce emissions listed of the designated TRI pollutants in two phases, with a 33 percent reduction by 1992 and a 50 percent reduction by 1995, thus giving the 33/50 name for the program (Kerret and Tal 2005). The 33/50 program was completely voluntary, and there was no enforcement mechanism to ensure that the reduction targets were met (Innes and Sam 2008). The main incentives for companies to participate in this program included to gain public recognition for pollution control efforts and to enhance a company's reputation with EPA (Coglianese and Nash 2008; Davies and Mazurek 1996).

By the end of the program, approximately 1,300 facilities were participating, and most tended to be large corporations (US EPA 1999; Coglianese and Nash 2008). Overall releases from both participating and non-participating companies declined 56% between 1988 and 1995, and the two-stage national reduction goals of 33 and 50% were met (Coglianese and Nash 2008). Despite meeting its stated goals, the 33/50 program is not credited as being the sole driver of the reduction; other factors that influenced the reduction included the use of 1988 as the baseline year to begin measuring emissions so that companies could get credit for work they began prior to 1991, and the fact that companies could eliminate the requirement to report emissions under the TRI program if they reduced their use of certain toxic chemicals below designated levels (Coglianese and Nash 2008; Kerret and Tal 2005). Additionally, EPA did not distinguish between reductions made by program participants and non-participants but measured the reduction in the aggregate (Davies and Mazurek 1996). The EPA's goal of encouraging reductions at the source also was met, as participating facilities reported approximately 30% more source reduction activity for 33/50 chemicals than for other TRI chemicals (US EPA 1999). Instead of mandating end-of-pipe controls, the 33/50 Program gave participants the freedom to pursue creative solutions, and it appears that companies did in fact pursue innovation.

It is difficult to evaluate the success of the 33/50 Program in a vacuum, as other factors impacted and contributed to results, as noted above. However, it does appear that the 33/50 Program demonstrated that flexibility – both in allowing an individual company to set their targets and then in determining how to accomplish those targets – was effective and helped offset the problem of regulations that require the maintenance of status quo technology. The program also demonstrated that a valuable, collective societal goal – a reduction in pollution – was not compromised by a voluntary program granting individual companies significant flexibility.

12.2.2 Common Sense Initiative

EPA's Common Sense Initiative (CSI), launched in 1994, was an industry-government collaborative effort to produce "cleaner, cheaper, and smarter" regulatory frameworks that would integrate environmental performance for an entire industry sector (Kerr et al. 1999). The initiative represented a shift in the agency's traditional focus of managing specific media and pollutants to a more holistic, cross-media, industry-wide approach (Davies and Mazurek 1999). In general, improved environmental protection was to be accomplished primarily by identifying regulatory requirements that created barriers to innovation in environmental technology and protection (Davies and Mazurek 1999). The EPA hoped that the collaborative effort would yield consensus as to how best to change the existing statutes and regulatory requirements in order to stimulate longer term capital investment in new technologies (Davies and Mazurek 1999). The goal was to provide incentives and flexibility to industry so that businesses would develop cost-effective, innovative technologies that either met or exceeded environmental standards (Fiorino 1996).

Six industries were included in the effort: automobile manufacturing, computer and electronics equipment, metal finishing, petroleum refining, printing, and steel (President's Council 1997). For each industry, representatives from business, environmental and community groups, labor organizations, and federal, state and local governments met as stakeholders to determine recommendations for changes to national environmental policies. Each industry team sent their analysis of issues and recommendations to a CSI Council composed of representatives from all stakeholder groups across all six industries. The CSI Council reviewed the teams' inputs and then made recommendations to EPA, which had final decision-making authority. The goal was to change the existing array of complicated, inconsistent policies into a comprehensive sector strategy (President's Council 1997).

CSI offered industry the possibility of reforming laws and regulations that were either redundant or imposed conflicting requirements, and CSI appeared to offer the potential to create flexible alternatives to current regulations, such as simplified reporting and record-keeping requirements and a streamlined permitting process (Davies and Mazurek 1997). However, despite the initiative's original promise, CSI is viewed as having limited success (Sousa and Klyza 2007; Caldart and Ashford

1999). In EPA's final evaluation of the initiative, the agency noted that only four projects out of forty led to recommended rule revisions that EPA acted upon, and that most CSI participants gradually came to believe that the initiative would not achieve "far-reaching change to EPA's rules and regulations" (Kerr et al. 1999).

Most criticism centers on the program's lack of substantive results due to high transaction costs and a lack of statutory authority on the part of EPA to grant regulatory waivers to industry participants. In terms of process barriers, the requirement to reach consensus was a major impediment, and the amount of time required for decision-making produced high transaction costs (Davies and Mazurek 1997). Environmental groups complained that they were underrepresented and several representatives from these groups resigned (Davies and Mazurek 1997). Additionally, some industry participants were uneasy sharing their proprietary information with either the government or their competitors, and some industry representatives feared that environmental groups might use the information to mount citizen lawsuits (Caldart and Ashford 1999). By 1996, two participants were labeled as obstructionist and were dismissed from CSI by EPA; two industries also ended participation in the initiative of their own accord, complaining about the onerous process (Davies and Mazurek 1997).

Perhaps the most significant issue with the CSI was the lack of statutory authority for either the CSI board or EPA to exempt regulated entities from existing regulations. The high transaction costs stem largely from this lack of statutory authority; when the government has no legal authority, it is driven to act by achieving some degree of consensus (Davies and Mazurek 1997). Also, results were hampered further by the lack of a pending regulatory hammer or penalty. Because the CSI committees functioned more as an advisory board than a direct participant in negotiated rulemaking, the impetus to develop a rule was less pronounced; the committee did not need to produce a rule before EPA did, so that the stakes were less defined than in some other projects, where if the group did not produce a specific rule, the agency would (Caldart and Ashford 1999). Another commonly cited criticism of the program is that it lacked a clearly defined mission: committees did not have a shared vision beyond "cleaner, cheaper, smarter" – and that vision was never defined (President's Council 1997). The CSI was officially terminated in 1998, but elements of the initiative were transitioned into EPA's Sector Strategies Program, which continues today (Kerr et al. 1999).

12.2.3 Project XL

Project XL (short for eXcellence and Leadership) was launched by EPA in 1995 as part of President Clinton's "Reinvention of Environmental Regulation" initiative (Lund 2000). A basic premise of Project XL was that the EPA's rigid, strict compliance system encouraged companies to simply follow the EPA's standards and discouraged investment in new technologies or approaches that could improve environmental performance beyond the current regulatory requirements (Caballero 1998). Unlike CSI, which focused more on reforming regulatory standards, XL focused on waiving enforcement in exchange for improved environmental performance. The regulated entity was allowed to propose an innovative means of achieving superior performance, and if approved, EPA suspended the traditional regulatory requirements (Hirsch 2001b). Under Project XL, companies signed a legally-binding contract in which they agreed to reduce pollution for a specific facility above what was required by law (Kerret and Tal 2005). In return, EPA engaged in site-specific rulemaking that implemented the regulatory waivers sought by the company (Hirsch 2001b). A common proposal was the establishment of "cap and trade" permits, which allowed facilities to make production changes to their operations without undergoing an agency review, as long as the company remained within their overall emissions limits (Wyeth 2006). Other projects included using technology to prevent pollution at its source instead of installing control equipment to the existing process, and using environmental management systems as a basis for consolidating permits (Lund 2000).

For example, Intel requested a "pre-approved" permit under Project XL for their Chandler, Arizona, semiconductor manufacturing plant which allowed the company to make process changes without needing to seek and obtain a revised permit from regulators (Davies and Mazurek 1997). The agreement granted Intel a facility-wide cap on air pollutant emissions, which eliminated the need for individual permits for different sources of air pollutants (U.S. EPA 1998). It is estimated that Intel's savings were in the millions, as a result of eliminating 30–50 regulatory reviews and requiring fewer permits (U.S. EPA 1998). Such a program provides significant benefits to a firm such as Intel operating in a "quick-to-market" industry (Davies and Mazurek 1997); the company developed a new generation of microprocessor every two to three years and yearly made between thirty and forty-five significant changes to its manufacturing process (Hirsch 2001b). In exchange for regulatory relief, Intel committed to maintain emissions levels at the site to a level defined as "minor" by the Clean Air Act, regardless of changes to the production process or whether a new manufacturing facility was built at the site (U.S. EPA 1998). Intel also made other commitments, such as implementing an environmental management system and reducing water consumption and the generation of both hazardous and non-hazardous waste (U.S. EPA 1998).

Both the Project XL program and the idea of regulatory flexibility proved controversial (Wyeth 2006). Some environmental groups viewed the regulatory waivers as a concession to industry and big business (Wyeth 2006). Critics claimed that the project violated the law because EPA did not have the authority to waive statutory requirements (Coglianese and Nash 2008). In fact, a common quote from the time was "if it isn't illegal, it isn't XL" (Coglianese and Nash 2008). XL produced some successes, resulting in approximately 40 final agreements (Sousa and Klyza 2007). However, compared to other EPA programs, participation in Project XL was limited to relatively few companies, with many firms choosing not to pursue innovative changes under the program due to the risk associated with the lack of statutory authority (Davies and Mazurek 1997). XL stopped accepting projects in 2003. There are three major issues that led to the closure of Project XL: (1) questionable legal authority for EPA to grant regulatory relief and protect participants from citizen lawsuits under existing environmental statutes; (2) lack of clarity around how the goal of "superior environmental performance" was defined and enforced; and (3) significant process barriers, such as the substantial time (average time 26 months) and cost (>$350,000) to negotiate an XL agreement (Caballero 1998; Coglianese and Nash 2008; Davies and Mazurek 1997).

12.2.4 Performance Track

Performance Track was launched in 2000 as part of EPA's effort to "reinvent" environmental regulation (OIG 2007). The program was intended to reward companies that achieved superior environmental performance (OIG 2007). In order to participate in the program, a facility had to complete a 22 page application that required extensive documentation of past achievements, a demonstrated record of sustained environmental compliance, and a commitment for specific future actions and achievements, including the commitment to improve environmental performance, to implement a formal environmental management system, and to engage in community outreach (Coglianese and Nash 2008). In exchange for making commitments for greater environmental protection, companies were offered relief from routine regulatory inspections, provided relief from some reporting and permitting requirements, given public recognition and favorable publicity, and provided networking and information exchange opportunities (Coglianese and Nash 2008).

The program produced mixed results. As with other initiatives, such as the 33/50 program, it is difficult to document whether the program alone is responsible for a reduction in environmental pollutants (Coglianese and Nash 2008). In 2006, a Harvard University study funded by EPA found that the prospect of membership in Performance Track did not necessarily motivate firms to improve their environmental performance, and that members' performance did not necessarily exceed the environmental performance of non-members (Coglianese and Nash 2008). A 2007 report from the Office of the Inspector General (OIG) reported both positive and negative criticism. While many participating firms had superior toxic release performance than their industry as a whole, a "substantial minority" performed worse than their industry counterparts (OIG 2007). Additionally, the report found that only 2 of 30 sampled program members "met all of their environmental improvement commitments." (OIG 2007). Performance Track was terminated in May 2009, after almost nine years of operation, at which time it had 547 members (US EPA 2009).

12.2.5 The Dutch Covenants Model

In general, the use of voluntary agreements in the United States tends toward site-specific "achievement" initiatives that allow flexibility to a regulated firm by gearing programs to fit the specific circumstances of the firm (Kerrret and Tal 2005). Additionally, the agreements usually are not legally binding and may lack an enforcement mechanism. In contrast, European countries often have utilized legally-binding, industry-wide "macro-contracts" that set specific performance standards, versus a general goal of "superior environmental achievement" as in the United States (Kerrret and Tal 2005). For example, the Netherlands has used a "covenant" model since the mid-1980s (OECD 2002). A covenant is a legally-binding, negotiated agreement between industry and government that specifies performance goals for the industry as a whole. In the "Dutch covenant" model, companies within an industry may choose whether to participate in an agreement, but once they voluntarily agree to the collectively-negotiated goals, they are then legally bound by the terms of the agreement (Harjula 1998).

In the Netherlands, the government developed the "Dutch covenant" model by working with selected industry sectors to set pollution reduction goals (Fiorino 1996). Industries were subdivided on the basis as to whether the sectors were "homogenous" or "heterogeneous". "Homogenous" sectors were characterized by companies that utilize similar operations and processes, and for those sectors, standards were negotiated for the sector as a whole. In "heterogeneous" sectors, processes are variable and complex, making it difficult to set sector-wide standards; for those sectors, agreements were negotiated with individual companies, and the individual company's goals fell within the overall sector goals (Fiorino 1996). Each company committed to achieving a negotiated share of the sector-wide pollution reductions. Thus the covenant became a plan for managing the environmental performance of an individual company as it fit within a sector or the sector as a whole, depending on whether the industry was homogenous or heterogeneous.

There are several benefits to a covenant approach. For example, because industry usually has more input in the development of a covenant than in traditional regulation, the covenants are usually "more practical and workable" for the industry (Hirsch 2006). Also, covenants often delineate performance goals instead of technology-based requirements, thus providing industry with flexibility in how to meet the goals. Technology is not prescribed. Additionally, covenants allow an industry to allocate a goal among sector participants so that those who can achieve the reduction most efficiently are allowed to do so; this mitigates some of the inefficiency inherent in traditional regulation (Hirsch 2006). In terms of innovation, covenants can act to either stimulate or restrict innovation. On one hand, covenants usually run for many years, and during this negotiated time period, the government may agree to maintain the established standards (Hirsch 2006; Fionori 1996). This allows companies to do long term planning that may incorporate capital investment and technological innovation. On the other hand, a covenant could remove the incentive to innovate, depending on the targets that are established.

12.3 Advantages and Disadvantages of Voluntary Collaborations

A review of the history of voluntary collaborative programs shows mixed results (Borck et al. 2008; Kerret and Tal 2005; Strasser 2008). Some of the projected benefits from voluntary programs were realized, but to a lesser degree than expected, and unanticipated negatives also resulted. A survey of the advantages and disadvantages of voluntary collaborative programs indicates there are important impacts on both the positive and negative sides of the ledger.

12.3.1 Strengths of the Collaborative Models

A key advantage of voluntary collaborative programs is the provision of flexibility. A collaborative agreement may allow businesses the flexibility to produce results better than the required legal minimum standard, so that the focus shifts from mere compliance to looking at how to generate continuous improvement over time (Wyeth 2006). Compliance may become merely the "starting point". More creative solutions are made possible by this approach, and this flexibility facilitates longer term capital planning, which includes investment in new technology (Kerret and Tal 2005). Flexibility was an underlying tenet of all programs used as case studies in this analysis, although it is difficult to quantify the benefits that resulted from this additional flexibility.

In addition, voluntary collaborative programs can promote cooperative relationships between business, the government, NGO's, communities, and interested citizen groups. A more cooperative relationship was expected to lead to faster, less expensive and more informed decision making with reduced transaction costs, and ultimately to improved environmental performance with fewer violations. While these results have been found largely lacking in the examples discussed above (Wyeth 2006; Gunningham 2009a), other benefits emerged. A collaboration can produce movement and consensus on an issue when political support for confronting an issue is lacking or when opposing political parties are at impasse, for example (Kerret and Tal 2005). Additionally, a collaboration with community or other special interest groups can generate good will leading to an enhanced corporate reputation for a company, which may be important to the long-term fiscal health of a business.

Another major expected advantage of voluntary collaborative programs was that better policies and strategies would result from greater information sharing. Such information sharing was an underlying tenet of all programs used as case studies, although the type of information sharing may have differed. In 33/50 and CSI, the government provided forums for industry participants to meet and exchange information. In addition, CSI convened multi-stakeholder meetings that supplied recommendations to EPA. In Performance Track, the government publicized the efforts of companies viewed as top performers in order to stimulate others to make similar environmental gains. While the limited analysis of voluntary programs has shown that such programs did not consistently produce the expected gains in environmental performance (Strasser 2008), few would doubt that in general, better decisions are made with better information, and most would agree that it is possible to improve policies if they are based on better and more complete information.

12.3.2 Weaknesses of the Collaborative Models

A number of reviews of the voluntary initiatives discussed above have found few or limited demonstrated benefits (Gunningham 2009a; Wyeth 2006; Kerret and Tal 2005). The central role that industry played in setting targets, poor monitoring of results, free riding by some companies, the uncertainty of regulatory threats and citizen suits created by a lack of statutory authority for EPA to grant regulatory waivers, and the fact that agreements were largely unenforceable due to a lack of sanctions or penalties, were often cited as reasons why results were less than expected (Gunningham 2009a). Additionally, there were concerns about the amount of time and resources required to produce a multi-stakeholder agreement, which led to high transaction costs (Hirsch 2001b; Caballero 1998). Finally, environmental advocates often looked cynically upon voluntary agreements, viewing them as "cosmetic attempts" by industry to appear as responsible corporate citizens, and categorized the agreements as "greenwash" due to the lack of sanctions associated with voluntary programs (Kerret and Tal 2005). Environmental groups often viewed collaborations as government-authorized "back-sliding" on environmental protection (Wyeth 2006).

12.4 Lessons Learned

From the case studies studied above, along with the scholarly commentary on these and other voluntary programs, several lessons can be distilled for the design and implementation of possible voluntary programs for emerging technologies. Perhaps the most basic lesson learned is that flexibility is necessary in the crafting of effective regulatory solutions involving voluntary programs. Traditional regulations are unable to anticipate all future scenarios and contexts in which they may be applied, especially for fast moving fields like emerging technologies (Hirsch 2001b). Voluntary collaborative agreements can be structured to adjust or "self-correct" over time, thus allowing firms to adapt more quickly to rapid changes in the industry without necessarily sacrificing the integrity of regulations.

A second basic lesson is that reward must at least approximate the assumed risk. If incentives are weak and transaction costs are too high, a program is less likely to succeed (Davies and Mazurek 1996). Quite simply, if a program is overly burdensome and has high transaction costs, participants are less likely to join. Participation is further threatened when an agency's lack of statutory authority to grant regulatory exceptions makes a program too risky. Third, voluntary programs tend to be more effective if backed-up by the threat of sanctions or enforcement, including the imposition of traditional regulatory requirements in the event of non-compliance with a voluntary program (Gunningham 2009a; Wyeth 2006). As one empirical analysis summarized its findings, "[t]he history of voluntarism would suggest that where the private interests of polluters in maintaining profitability and the public interest in protecting the environment do not substantially coincide, then (unless there are countervailing economic or social pressures) pure voluntarism will be largely ineffective in changing behavior" (Gunningham 2009b, p. 161).

Fourth, an effective agreement has ambitious, clearly defined goals and a mechanism for amending goals over time, along with an effective monitoring and sanctions system. This helps facilitate and sustain agreement between varied constituencies over time. Also, measurable outcomes are necessary for determining success. Fifth, involvement of varied constituencies tends to create an incentive for companies to comply with the agreements, as such involvement adds transparency and improves the prospect for public support (Kerret and Tal 2005). Finally, a successful collaborative process requires a significant investment in resources, particularly in staff time allocated to the effort. The procurement cycle for each participating entity should be considered, as delays can sabotage the long-term success of a negotiation, and participants can walk away after incurring significant transaction costs. Parties to such a collaborative process should be made aware of and plan for the necessary investment of time and resources, otherwise, the effort may dissipate over time.

12.5 Conclusion

The success of recent voluntary collaborative programs in the environmental field is mixed. Some scholars consider the focus on voluntary and negotiated agreements to be largely unsuccessful, as voluntary performance standards have not been consistently associated with improved environmental performance (Gunningham 2009a; Strasser 2008). However, direct and indirect benefits have been obtained under at least some voluntary programs. The OECD concluded that properly designed voluntary programs can "play a useful role in 'lubricating' [the] policy mix; increasing flexibility, paving the way for new regulations without a stringent and brutal implementation, inducing industry to develop innovative approaches, [and] filling enforcement deficits. . . ." (OECD 2000). Thus, voluntary programs remain a viable tool to consider for oversight of emerging technologies, especially as an interim measure when no traditional regulation exists or is feasible (Marchant et al. 2008).

There is ample information to be gleaned from recent regulatory experiments so that we are now better positioned to determine which mechanism will work best in a particular industry at a particular time. If the required statutory authority is granted to a rulemaking agency, then regulation can be used effectively as either an incentive or a penalty to supplement a voluntary collaborative program that is tailored to fit a given situation. For some collaborations, an industry sector approach may be most effective; at other times, a site-specific approach may be optimal. Regardless, flexibility is critical. Otherwise, regulations are likely to continue to fall behind the pace of science and technology and will continue to exact a high societal cost.

References

Alberini, Anna and Kathleen Segerson. 2002. Assessing voluntary programs to improve environmental quality. *Environmental and Resource Economics* 22: 157–184.

ASTM International Nanotechnology Standards available at http://www.astm.org/Standards/nanotechnology-standards.html; ISO Standards developed by the ISO's Technical Committee 229 on Nanotechnologies is available at http://www.iso.org/iso/iso_catalogue/catalogue_tc/catalogue_tc_browse.htm?commid=381983&includesc=true&published=on&development=on.

Borck, Jonathan C., Cary Coglianese, and Jennifer Nash. 2008. Environmental leadership programs: Toward an empirical assessment of their performance. *Ecology Law Quartely* 35: 711–833.

Breggin, L.K., and L. Carothers. 2006. Governing uncertainty: The nanotechnology environmental, health, and safety challenge. *Columbia Journal of Environmental Law* 31: 285–329.

Caballero, Thomas E. 1998. Project XL: Making it legal, making it work. *Stanford Environmental Law Journal* 17: 399–471.

Caldart Charles C., and Nicholas A. Ashford. 1999. Negotiation as a means of developing and implementing environmental and occupational health and safety policy. *Harvard Environmental Law Review* 23: 141–202.

Coglianese, Cary, and Jennifer Nash. 2008. Government clubs: Theory and evidence from voluntary environmental programs. *University of Pennsylvania Law School*, 6–7. Public law and legal theory research paper series, research paper no. #08–49, available at http://ssrn.com/abstract=1311340.

Commission of the European Communities (CEC). 2008. *Commission recommendation on a code of conduct for responsible nanosciences and nanotechnologies research*, dated 7/2/2008, available at: http://ec.europa.eu/nanotechnology/pdf/nanocode-rec_pe0894c_en.pdf.

Davies, Terry, and Jan Mazurek. 1996. *Industry incentives for environmental improvement: Evaluation of U. S. federal initiatives*. Washington, DC: Global Environmental Management Initiative.

Fiorino, Daniel J. 1996. Toward a new system of environmental regulation: The case for an industry sector approach. *Environmental Law* 26: 457–488.

Gunningham, Neil. 2009a. Environmental law, regulation and governance: Shifting architectures. *Journal of Environmental Law* 21: 179–212.

Gunningham, Neil. 2009b. The new collaborative environmental governance: The localization of regulation. *Journal of Law and Society* 36: 145–166.

Harjula, Henrik. 1998. *Extended producer responsibility, phase 2, case study on the Dutch packaging covenant*. Organisation for Economic Co-operation and Development, 15 May 1998.

Hirsch, Dennis D. 2006. Protecting the inner environment: What privacy regulation can learn from environmental law. *Georgia Law Review* 41: 1–64.

Hirsch, Dennis D. 2001a. Second generation policy and the new economy. *Capital University Law Review* 29: 1–20.

Hirsch, Dennis D. 2001b. Project XL and the special case: The EPA's untold success story. *Columbia Journal of Environmental Law* 26: 219–257.

Innes, Robert, and Abdoul G. Sam. 2008. Voluntary pollution reductions and the enforcement of environmental law: An empirical study of the 33/50 program. *Journal of Law and Economics* 51: 271–296.

Isenberg, Howard. 1995. The second industrial revolution: The impact of the information explosion. *Industrial Engineering* 27: 15.

Johnson, Stephen M. 2001. Economics v. equity II: The European experience. *Washington and Lee Law Review* 58: 417–486.

Karkkainen, Bradley C. 2006. Information-forcing environmental regulation. *Florida State University Law Review* 33: 861–902.

Kerr, Greiner, Anderson, and April, Inc. 1999. Analysis and evaluation of the EPA's common sense initiative, http://www.epa.gov/evaluate/pdf/pubs_finalcsi.pdf.

Kerret, Dorit, and Alon, Tal. 2005. Greenwash or green gain? Predicting the success and evaluating the effectiveness of environmental voluntary agreements. *Penn State Environmental Law Review* 14: 31–84.

Litan, Robert E. 2001. Law and policy in the age of the internet. *Duke Law Journal* 50: 1045–1085.

Lund, Lisa C. 2000. Project XL: Good for the environment, good for business, good for communities. *Environmental Law Review* 30: 10140–10152.

Marchant, Gary E., Douglas S. Sylvester, and Kenneth W. Abbott. 2008. Risk management principles for nanotechnology. *NanoEthics* 2: 43–60.

Occupational Safety & Health Administration. 2007. All about VPP, available at http://www.osha.gov/dcsp/vpp/sitebysic.html

Office of Inspector General, EPA. 2007. Evaluation report: Performance Track could improve program design and management to ensure value, available at http://www.epa/gov/oig/reports/2007/2007/20070329-2007-P-00013.pdf.

Organisation for Economic Cooperation and Development (OECD). 2002. *Regulating policies in OECD countries: From interventionism to regulating governance – Annex II*, OECD Reviews of Regulatory Reform: Regulatory Policies in OECD Countries.

Organisation for Economic Cooperation and Development (OECD). 2000. *Voluntary approaches for environmental policy: An assessment*. Paris: OECD Publishing.

Pearson, Eric. 2002. Environmental and Natural Resources Law. Albany, NY: LexisNexis Matthew Bender, 2d ed.

President's Council on Sustainable Development, New National Opportunities Task Force. 1997. Lessons learned from collaborative approaches, available at http://clinton2.nara.gov/PCSD/Publications/Lessons_Learned.html.

Sousa, David J. and Christopher McGrory Klyza. 2007. New directions in environmental policy making: An emerging collaborative regime or reinventing interest group liberalism? *Natural Resources Journal* 47: 377–444.

Strasser, Kurt A. 2008. Do voluntary corporate efforts improve environmental performance?: The empirical literature. *Boston College Environmetal Affairs Law Review* 35: 533–555.

U.S. Environmental Protection Agency. 2009. National Environmental Performance Track, memorandum from EPA administrator Lisa P. Jackson dated 16 March 2009. *Next steps for the national environmental Performance Track program and the future of environmental leadership programs*, available at http://epa.gov/performancetrack/downloads/PerformanceTrackNextStepsMemoExternal-text.pdf.

U.S. Environmental Protection Agency (EPA). 1999. 33/50 program: The final record. EPA-745-R-99-004, available at http://www.epa.gov/oppt/3350/3350-fnl.pdf.

U.S. Environmental Protection Agency (EPA). 1998. Intel XL project fact sheet, available at http://www.epa.gov/ProjectXL/intel/09998.html.

U.S. Environmental Protection Agency. 2009. EPA partnership programs: List of programs, available at http://www.epa.gov/partners/programs/index.htm.

Wyeth, George B. 2006. "Standard" and "Alternative" environmental protection: The changing role of environmental agencies. *William & Mary Environmental Law and Policy Review* 31: 5–73.

Zinn, Matthew D. 2002. Policing environmental regulatory enforcement: Cooperation, capture, and citizen suits. *Stanford Environmental Law Journal* 21: 81–174.

Part IV
Conclusion

Chapter 13
Addressing the Pacing Problem

Gary E. Marchant

The central conclusion from the cumulative insights of the contributions to this volume is that existing regulatory systems and ethical frameworks are inadequate to provide effective, meaningful and timely oversight of the current and future generations of emerging technologies. Technologies such as genetics, robotics, information technologies, nanotechnology, synthetic biology, and neuroscience are racing forward at a pace of technology development that has never before been experienced in human history. In contrast, our traditional government oversight systems are mired in stagnation, ossification and bureaucratic inertia, and are seriously and increasingly lagging behind the new technologies accelerating into the future.

The challenge to law and ethics to keep pace with rapidly developing emerging technologies is affected by other dynamics in addition to the disparity of the relative speeds of the two domains. The oversight of emerging technologies is more complex than most previous regulatory challenges, in that the technologies involved tend to have many diverse applications and forms, are used in many different industries and contexts, and present a multitude of different and often hard-to-quantify risk and benefit scenarios. These technologies often fail to fit comfortably within existing regulatory categories, and thus the path dependency created by regulatory frameworks developed for earlier, simpler technologies are problematic (Moses 2007). Emerging technologies often present important social, ethical and privacy risks beyond comparatively straightforward health, safety and environmental risks, which our regulatory systems are poorly constructed to address (Marchant et al. 2010). These technologies are being developed in a new era of public scrutiny and increased role for NGOs, requiring better forms of public engagement that do not yet exist. The lessons from past debacles such as asbestos and chloroflourocarbons (CFCs) have created an imperative to address anticipated rather than known risks, which further stretches our risk assessment and risk management capabilities. These challenges further exacerbate the growing asynchronicity in the pacing of law and ethics with science and technology.

G.E. Marchant (✉)
Arizona State University, Tempe, AZ, USA
e-mail: gary.marchant@asu.edu

G.E. Marchant et al. (eds.), *The Growing Gap Between Emerging Technologies and Legal-Ethical Oversight*, The International Library of Ethics, Law and Technology 7,
DOI 10.1007/978-94-007-1356-7_13, © Springer Science+Business Media B.V. 2011

If only to call attention to this obvious but under-studied “Pacing Problem,” this book has served an important purpose. But what then is to be done? There are two possible strategies to attack this problem, which are not necessarily mutually exclusive. One strategy would be to try to slow the pace of technology development so that law and ethics are less likely to lag behind the developing technologies. The second strategy would be to try to accelerate the adaptivity and responsiveness of law and ethics to better keep up with advancing technology.

The first strategy of trying to slow technology is represented by the precautionary principle. The precautionary principle, often encapsulated by the phrase “better safe than sorry,” requires regulators to err on the side of safety by delaying new technologies until their safety can be adequately ensured (Raffensperger and Tickner 1999). Historical support for such an approach is provided by a EU-commissioned study entitled *Late Lessons from Early Warnings* that documented fourteen examples, including asbestos, chlorofluorcarbons (CFCs), and lead in gasoline, where governments and industry failed to adequately heed indications of an unacceptable health risk, to the ultimate detriment of public health and often the very industry promoting the technology at issue (Harremoës et al. 2001). To avoid repeating such mistakes, the precautionary principle seeks to create a “speed bump” that can slow the pace of rapidly developing technologies whose risks are uncertain and regulatory frameworks incomplete. As such, the precautionary principle may be seen as having considerable promise for addressing the Pacing Problem.

Yet, there are several limitations of the precautionary principle that suggest it may not be an appropriate or effective solution to the Pacing Problem, at least single-handedly. The precautionary principle has been controversial in the international arena, with the European Union being its strongest advocate while the United States has been more skeptical, perhaps due to the highly legalistic American regulatory environment. One focus of the controversy has been that the precautionary principle lacks a explicit or consensus definition, with dozens of different formulations offered with subtle differences in wording that translate into potentially significant differences in application (Sandin 1999). This ambiguity opens the precautionary principle to criticism of inconsistent or arbitrary application (Marchant and Mossman 2004). Moreover, none of the versions offered to date answer critical questions inherent in regulatory decisions such as what evidence of harm is sufficient to trigger precautionary action, what type of evidence of safety must a manufacturer produce to satisfy its burden of proof, what level of risk is acceptable, and how should the costs of regulation and the benefits of technologies be weighed against its risks (if at all) (Marchant 2003).

This ambiguity of the precautionary principle is particularly critical when applied to emerging technologies that offer both potential health and environmental benefits *and* risks. In at least some cases, slowing new technologies in favor of maintaining the status quo with existing technologies may have the net effect of blocking health and environmental improvements (Cross 1996; Holm and Harris 1999). For example, while nanotechnology undoubtedly presents some real but uncertain risks, nanotechnology also offers many promising health and environmental benefits, including more effective and safer cancer treatments, improved medical diagnostics, remediation of hazardous wastes, cleaner energy technologies, and improved control of pollution emissions (EPA 2007; Gwinn and Vallyathan 2006). It is quite possible that a moratorium or delay of nanotechnology development pursuant to the precautionary principle would do more harm than good to human health and the environment. In such situations, plausible arguments could be made, depending on one's assumptions and risk-benefit estimates, that the precautionary principle simultaneously supports both restricting and promoting the same technology (Sunstein 2003).

To be sure, some reasonable precaution can and should be an important part of the governance of emerging technologies. It is in the common interest of all concerned, including government, industry, civil society, and the general public, to try to prevent proactively significant harms from emerging technologies which will not only be contrary to public well-being, but could also block further development of the relevant technology, with the attendant loss of potential benefits and existing investments. There is therefore both room and a need to develop sensible and realistic models for prudent application of precaution to emerging technologies. But given the powerful economic and technological drivers propelling the emerging technologies forward, as well as the enormous benefits they portend, it would be both unlikely and likely misguided to try to use precaution to block the progress of these technologies outright. Accordingly, there will continue to be a need for the second possible strategy for addressing the Pacing Problem, of trying to speed up the development and adaptation of legal and ethical frameworks to address accelerating technologies.

Addressing the Pacing Problem by developing strategies and solutions to make law, policy and ethics better adept at keeping pace with accelerating science and technology will not be an easy chore. As one long-time government expert has recently written:

> The increasing rapid pace of technology change of all kinds presents modern societies with some of their most pressing challenges. Rapid change demands foresight, vision, adaptability, and creativity, all combined with a healthy degree of prudence. Such capabilities are difficult to come by in the complicated and often messy world of modern governance (Fiorino 2010)

While appropriately daunted and humbled by the challenge, this volume has made an initial effort to not only document and bring attention to the Pacing Problem, but has also made an initial attempt to explore possible approaches and mechanisms to address this problem. While far more work needs to be done in developing solutions, the initial conclusions from this project are that the Pacing Problem must be addressed by some combination of (i) adaptive governance, (ii) soft law mechanisms, and (iii) institutional reform.

Adaptive Governance: Given the pace at which emerging technologies are being developed, it simply is not feasible to regulate these technologies adequately and comprehensively at the "front end," before we can possibly know the future direction of the technologies and the problems they may present (Shapiro and Glicksman 2003). Accordingly, it is critical to implement processes that permit frequent and ongoing reevaluation and revisions of regulatory programs to address changing facts and circumstances. Adaptive governance derives from the concept of adaptive management first developed in the context of ecology to experiment with different policy approaches that are simultaneously undertaken with active monitoring, assessment and adjustment (Holling 1978; Ruhl 2005). This iterative process requires both active monitoring to detect relevant changes as soon as possible, and the capability to recalibrate oversight requirements quickly in response to such changes (Garmestani et al. 2009; Ruhl 2005). As the International Risk Governance Council (IRGC) stated in recommending an adaptive approach in the oversight of emerging technologies such as nanotechnology, such an approach requires "valuing flexibility in the application of risk management strategies as knowledge and understanding of the field develops" (IRGC 2007).

In other words, regulatory and oversight systems must be designed from the outset to expect, anticipate and be able to respond to change. This is the key concept of adaptive governance, as explained by Nobel Prize winning economist Elinor Ostrom and her colleagues:

> Institutions must be designed to allow for adaptation because some current understanding is likely to be wrong, the required scale of organization can shift, and biophysical and social systems change. Fixed rules are likely to fail because they place too much confidence in the current state of knowledge, whereas systems that guard against the low probability, high consequence possibilities and allow for change may be suboptimal in the short run but prove wiser in the long run. This is a principal lesson of adaptive management research (Dietz et al. 2003, p. 1909)

Yet, traditional administrative law generally acts to prevent rapid agency changes in position by requiring policy to be locked-in as "final" rules that can only be changed by going through time-consuming and burdensome rulemaking procedures. In the words of one commentator: "[T]he clear message to agencies under conventional administrative law is that they adopt adaptive management at their own peril. Adopting adaptive management may be an agency's dream; practicing it is a nightmare" (Ruhl 2005, p. 39). New mechanisms are therefore needed to circumvent or overcome such procedural impediments to adaptive management.

One mechanism for such adaptive management, proposed in the Chapter 9 by Kenneth Abbott (this volume), is the creation of a framework convention to govern emerging technologies, perhaps named the Framework Agreement on Scientific and Technological Innovation and Regulation (FASTIR). A framework convention has many of the important attributes of adaptive management in that it creates initially an institutional structure consisting of an organizing body, a regular process for the parties to meet, and some core principles, but is otherwise an empty shell originally that can gradually and incrementally be built up with substantive content in the form of protocols. In other words, the treaty is intended to evolve over time, with more substance added gradually in response to both growing knowledge and political will to take action. A framework convention also has the additional advantage of being international, an important attribute (yet one that at the same times introduces additional complexity and challenges) given that emerging technologies are in large part an international development.

Another factor for allowing for adaptive management is careful consideration of whether or not to adopt *sui generis* rules for new technologies, which as discussed in the Chapter 6 by Lyria Bennett Moses (this volume), can sometimes freeze into place regulatory structures relative to more broadly written legal frameworks that can better adapt to changes in technologies and the regulatory issues they present. Nevertheless, if *sui generis* laws are deemed necessary, they should be drafted to allow for the likelihood of future technology changes that may not be in anticipated directions (Moses, Chapter 6, this volume). Other adaptive management approaches suggested in this book includes anticipatory technology assessment proposed by Daniel Sarewitz (Chapter 7, this volume), an early warning system and list of known unknowns as proposed herein by David Rejeski (Chapter 4, this volume), and sunset and temporary provisions in regulations and statutes as proposed by Gaudet and Marchant (Chapter 11, this volume).

"Soft Law" Approaches: For many technologies, it will be necessary to rely (at lease initially) on voluntary or "soft law" approaches. Although originally developed in the international law context, soft law approaches are increasingly being used in national and local oversight programs (Gersen and Posner 2008; Marchant et al. 2008). As one set of authors recently noted, with the rapid pace of development of emerging technologies such as nanotechnology, "[i]t is likely that the complexity of the issues and the rapid pace of development will outstrip the capacities of the regulatory agencies to frame effective policies and standards. In such a scenario, reliance on responsible corporate behavior becomes a dominant rather than a transient consideration in devising regulatory structures." (Lee and Jose 2008, p. 117).

"Soft law" approaches involve a variety of instruments that establish substantive goals or norms that are not directly enforceable (Abbott and Snidal 2000). The related concept of "governance" approaches expand the responsibility for oversight from government exclusively to a broader range of stakeholders and actors including businesses, non-governmental organizations, various forms of partnerships and collaborations, networks, third party auditors and other entities. Soft law/governance approaches to oversight offer a number of potential advantages, including: (i) they are usually based on cooperative rather than adversarial models of engagement; (ii) they can be adopted or revised relatively quickly; (iii) many different soft law/governance concepts can be attempted simultaneously; and (iv) such measures can be gradually "hardened" into more formal regulatory instruments (Abbott and Snidal 2000; Gersen and Posner 2008).

A variety of "soft law" approaches for governance of emerging technologies have been considered in this volume. Rappert (Chapter 8, this volume) evaluates the role of codes of conduct, finding they have yet to have much beneficial effect in their actual application, but may provide a useful educational and expressive function in the drafting process for a code of conduct. Abbott's (Chapter 9, this volume) proposal herein for a Framework Agreement on Scientific and Technological Innovation and Regulation incorporates many elements of soft law and governance. Waugh and Marchant (Chapter 12, this volume) examine the role of voluntary agreements, perhaps the largest category of soft law instruments. The principles-based regulation analyzed by Carter and Marchant (Chapter 10, this volume) represents a hybrid or intermediary between traditional hard law and more recent soft law approaches.

Institutional Reform: Various types of institutional reforms can also help to address the Pacing Problem. This can consist of structural changes within existing agencies, such as creating a safety reporting system for reporting and studying errors, or embedding an "early warning officer" and support staff within agencies to scan the horizon for approaching issues and challenges, as David Rejeski (Chapter 4, this volume) proposed in his chapter in this volume. In the end, some new form of regulatory agency may be needed. For example, Furger and Fukuyama (2007) have called for a new regulatory institution designed to address the unique set of regulatory issues presented by emerging technologies in the life sciences. Various other commentators have suggested the need for a new institution to govern new technologies that is more adaptive, flexible and able to address the full range of issues presented by new technologies – including health, environmental, social, ethical and economic impacts (IRGC 2007; Davies 2008; FramingNano Project 2010; Marchant et al. 2010, 2009).

These three inter-related approaches of adaptive management, soft law, and institutional reform, along with the variety of legal mechanisms and proposals they encompass that are developed more fully in the various chapters in this volume, provide a set of tools to begin to address the Pacing Problem. But in addition to these tools, perhaps what is most needed is an awareness and reflexivity on behalf of law makers, policy makers, regulators, jurists, and scholars of the existence and urgency of the Pacing Problem, and the need to address it and its symptoms in a methodical and coherent manner (Moses 2007). Hopefully this monograph has contributed to that undertaking.

References

Abbott, Kenneth W., and Snidal, Duncan. 2000. Hard and soft law in international governance. *International Organizaion* 54: 421–456.

Cross, F.B. 1996. Paradoxical perils of the precautionary principle. *Washington and Lee Law Review* 53: 851–925.

Davies, J. Clarence. 2008. Nanotechnology oversight: An agenda for the new administration. Woodrow Wilson International Center for Scholars Project on Emerging Technologies.

Dietz, Thomas, Elinor Ostrom, and Paul C. Stern. 2003. The struggle to govern the commons. *Science* 302: 1907–1912.

Fiorino, Daniel J. 2010. Nanoscale regulation (letter). *Issues in Science and Technology* (Winter): 10–12.

FramingNano Project. 2010. The FramingNano governance platform: A new integrated approach to the responsible development of nanotechnologies, Final Report.

Furger, Franco, & Francis, Fukuyama. 2007. Beyond bioethics: A proposal for modernizing the regulation of human biotechnologies. *Innovations* (Fall): 117–127.

Garmestani, Ahjond S., Craig R. Allen, and Heriberto Cabezas. 2009. Panarchy, adaptive management and governance: Policy options for building resilience. *Nebraska Law Review* 87: 1036–1054.

Gersen, Jacob E., and Eric A. Posner. 2008. Soft law: Lessons from congressional practice. *Stanford Law Review* 61: 573–627.

Gwinn, M.R., and V. Vallyathan. 2006. Nanoparticles: health effects – pros and cons. *Environmental Health Perspectives* 114: 1818–1825.

Harremoës, P., D. Gee, M. MacGarvin, A. Stirling, J. Keys, B. Wynne, and S.G. Vaz. 2001. *Late lessons from early warnings: The precautionary principle 1896–2000*. European Environmental Agency Environmental Issue Report No. 22, available at http://www.eea.europa.eu/publications/environmental_issue_report_2001_22/Issue_Report_No_22.pdf.

Holling, C.S., ed. 1978. *Adaptive environmental assessment and management*. New York: Wiley.

Holm, S., and J. Harris. 1999. Precautionary principle stifles discovery (letter). *Nature* 400: 398.

International Risk Governance Council (IRGC). 2007. *Nanotechnology risk governance*. Geneva: IRGC.

Lee, Robert Lee, and P.D. Jose. 2008. Self-interest, self-restraint and corporate responsibility for nanotechnologies: Emerging dilemmas for modern managers. *Technology Analysis and Strategic Management* 20: 113–125.

Marchant, G.E. 2003. From general policy to legal rule: The aspirations and limitations of the precautionary principle. *Environmental Health Perspectives* 111: 1799–1803.

Marchant, G.E., and K.L. Mossman. 2004. *Arbitrary and capricious: The precautionary principle in the European Union Courts*. Washington: AEI Press.

Marchant, G., A. Meyer, and M. Scanlon. 2010. Integrating social and ethical concerns into regulatory decision-making for emerging technologies. *Minnesota Journal Law Science and Technology* 11: 345–363.

Marchant, Gary E., Douglas J. Sylvester, and Kenneth W. Abbott. 2009. What does the history of technology regulation teach us about nano oversight. *Journal Law, Medicine and Ethics* 37: 724–731.

Marchant, Gary E., Douglas J. Sylvester, and Kenneth W. Abbott. 2008. Risk management principles for nanotechnology. *NanoEthics* 2: 43–60.

Moses, Lyria Bennett. 2007. Recurring dilemmas: The law's race to keep up with technological change. *University of Illinois Journal of Law, Technology and Policy* 2007: 239–285.

Raffensperger, C., and J. Tickner, eds. 1999. *Protecting public health & the environment: Implementing the precautionary principle*. Washington, DC: Island Press.

Ruhl, J.B. 2005. Regulation by adaptive management – Is it possible? *Minnesota Journal of Law, Science and Technology* 7: 21–57.

Sandin, P. 1999. Dimensions of the precautionary principle. *Human and Ecological Risk Assessment* 5: 889–907.

Shapiro, Sidney A., and Robert L. Glicksman. 2003. *Risk regulation at risk: Restoring a pragmatic approach.* Stanford: Stanford University Press

Sunstein, C.R. 2003. Beyond the precautionary principle. *University of Pennsylvania Law Review* 151: 1003–1058.

U.S. Environmental Protection Agency (EPA). 2007. Science policy council, nanotechnology white paper, EPA 100/B-07/001, available at http://es.epa.gov/ncer/nano/publications/whitepaper12022005.pdf.

Index

G.E. Marchant et al. (eds.), *The Growing Gap Between Emerging Technologies and Legal-Ethical Oversight*, The International Library of Ethics, Law and Technology 7, DOI 10.1007/978-94-007-1356-7,

H

I

J

K

L

MIX
Papier aus verantwortungsvollen Quellen
Paper from responsible sources
FSC® C105338

If you have any concerns about our products,
you can contact us on
ProductSafety@springernature.com

In case Publisher is established outside the EU,
the EU authorized representative is:
Springer Nature Customer Service Center GmbH
Europaplatz 3, 69115 Heidelberg, Germany

Printed by Libri Plureos GmbH
in Hamburg, Germany